T0338522

Solved Problems in Transport Phenomena
Momentum Transfer

Solved Problems in Transport Phenomena

Momentum Transfer

İsmail Tosun

Middle East Technical University, Turkey

NEW JERSEY · LONDON · SINGAPORE · BEIJING · SHANGHAI · HONG KONG · TAIPEI · CHENNAI · TOKYO

Published by

World Scientific Publishing Co. Pte. Ltd.

5 Toh Tuck Link, Singapore 596224

USA office: 27 Warren Street, Suite 401-402, Hackensack, NJ 07601

UK office: 57 Shelton Street, Covent Garden, London WC2H 9HE

Library of Congress Cataloging-in-Publication Data

Names: Tosun, İsmail, author.

Title: Solved problems in transport phenomena : momentum transfer /
 İsmail Tosun, Middle East Technical University, Turkey.

Description: Hackensack, New Jersey: World Scientific, [2023] |
 Includes bibliographical references and index.

Identifiers: LCCN 2022020921 | ISBN 9789811256240 (hardcover) |
 ISBN 9789811256257 (ebook for institutions) | ISBN 9789811256264 (ebook for individuals)

Subjects: LCSH: Momentum transfer--Mathematical models--Textbooks.

Classification: LCC TP156.T7 T675 2023 | DDC 660/.284292--dc23/eng/20220713

LC record available at https://lccn.loc.gov/2022020921

British Library Cataloguing-in-Publication Data

A catalogue record for this book is available from the British Library.

For any available supplementary material, please visit
https://www.worldscientific.com/worldscibooks/10.1142/12834#t=suppl

Desk Editors: Balasubramanian Shanmugam/Steven Patt

Typeset by Stallion Press
Email: enquiries@stallionpress.com

Printed in Singapore

To my precious first grandson
Deniz Zachary Bashor
You are a blessing and a joy

Preface

Transport phenomena is an umbrella term to describe the fundamental processes of momentum, energy, and mass transfer. The publication of the world-renowned book *Transport Phenomena* by R. B. Bird, W. E. Stewart, and E. N. Lightfoot (University of Wisconsin-Madison professors) in 1960 caused a paradigm shift in chemical engineering education. Since then, many authors have published various books on the subject. However, from the students' point of view, the course on transport phenomena is still extremely difficult since it requires blending physical principles with mathematical vigor.

The purpose of this book is not to provide students with a large repertoire of problems on momentum transfer. Instead, the purpose is to teach the intricacies in the three stages of problem-solving, namely formulation, simplification, and mathematical solution. The problems selected for this purpose are not original at all; in fact, they can be found in textbooks as either examples or end-of-chapter problems. However, the methodology used in problem-solving is different and the presentation helps students to gain physical intuition and mathematical skills.

In fluid flow problems involving simple geometries, students usually take everything for granted. For example, laminar flow under steady conditions through a circular pipe is a classical problem solved in transport phenomena courses. Students solve the problem by assuming $P = P(z)$ and $v_z = v_z(r)$ but never question the reasoning behind it. The difficulty arises in flow through more complex geometries. My emphasis, therefore, is to show the students how to tackle a problem as if they were going to solve it for the first time in their lives.

This book covers momentum transfer at the microscopic and macroscopic levels. It can be used in transport phenomena courses in conjunction with standard textbooks. Although it is written for students majoring in chemical engineering, it can also be used as a supplementary text in environmental, mechanical, petroleum, and civil engineering courses related to fluid mechanics.

A basic knowledge of vector and tensor algebra is a prerequisite to understand the subject. Therefore, I strongly suggest the readers to go over Appendix A before they start reading Chapter 1.

My colleagues Emre Büküşoğlu, Yusuf Uludağ, and Halil Kalıpçılar made many helpful suggestions during the preparation of this book. Their efforts are greatly appreciated. Above all, I am grateful to my wife, Ayşe, for her unwavering support and understanding since the day we got married.

İsmail Tosun
(itosun@metu.edu.tr)

About the Author

 **Ismail Tosun** received his BS and MS degrees from the Middle East Technical University (METU) and a PhD degree from the University of Akron, all in chemical engineering. His research interests are mathematical modeling, solid–liquid separation processes, and multiphase transport phenomena. Currently, he is an emeritus professor of chemical engineering at METU. Professor Tosun is the author of the following books:

1. *Modeling in Transport Phenomena – A Conceptual Approach*, 2nd ed., Elsevier, 2007.
2. *Fundamental Mass Transfer Concepts in Engineering Applications*, CRC Press, 2019.
3. *Thermodynamics – Principles and Applications*, 2nd ed., World Scientific, 2020.
4. *The Thermodynamics of Phase and Reaction Equilibria*, 2nd ed., Elsevier, 2021.

Contents

Chapter 1

Introduction

1.1 Definitions

1.1.1 *Steady-state, uniform, and equilibrium*

It is important to differentiate between the concepts of steady-state, uniform, and equilibrium:

- **Steady-state:** The term "steady-state" means that at a particular location in space the dependent variable φ does not change as a function of time, i.e.,

$$\left(\frac{\partial \varphi}{\partial t}\right)_{x,y,z} = 0. \tag{1.1-1}$$

- **Uniform:** The term "uniform" means that at a particular instant in time the dependent variable φ is not a function of position, i.e.,

$$\left(\frac{\partial \varphi}{\partial x}\right)_{y,z,t} = \left(\frac{\partial \varphi}{\partial y}\right)_{x,z,t} = \left(\frac{\partial \varphi}{\partial z}\right)_{x,y,t} = \nabla \varphi = 0. \tag{1.1-2}$$

- **Equilibrium:** A system is in equilibrium if both the steady-state and uniform conditions are met simultaneously. This implies that the variables associated with the system, such as temperature, pressure, and density, are constant at all times and have the same magnitude at all positions within the system. A difference in any potential that causes a process to take place spontaneously is called a *driving force*. The driving force(s) turns out to be zero for a system in equilibrium.

1

1.1.2   *Kinematics*

Kinematics is the branch of mechanics that deals with motions of bodies. Two types of volume elements are considered in kinematics. An *Eulerian* (or *spatial*) volume element is fixed in space. Since the boundaries do not move, while its volume is constant, the mass contained in it is dependent on time. A *material* (or *Lagrangian*) volume element moves with the fluid velocity; its boundaries expand or contract so as to keep the mass contained in it constant.

A *path line* is the curve or path followed by a particle during flow. A *streamline* is the curve whose tangent at any point is in the direction of the velocity at that point, i.e., the velocity vector is everywhere tangent to the streamline. Path lines and streamlines become identical when the flow is steady.

1.1.3   *Time derivatives*

- **Partial time derivative:** The term "$\partial \varphi / \partial t$" is called the *partial time derivative* and represents the variation in φ with time at a fixed position in space. For example, consider a fluid flowing in a pipe. When we drill a hole and dip a thermometer into the pipe to record the temperature of the fluid as a function of time, we observe how the temperature changes with time at a fixed position in space, i.e.,

$$\left(\frac{dT}{dt} \right)_{\mathbf{r}} = \frac{\partial T}{\partial t}, \tag{1.1-3}$$

  where $\mathbf{r}$ is the *spatial position vector* locating the fluid particle in space.
- **Material (substantial) time derivative:** The term "$D\varphi / Dt$" is called the *material time derivative* and represents the variation in φ with time that occurs when an observer moves with the velocity of the fluid. For example, when we place a saddle on a fluid particle (hypothetically, of course!) and ride on it with a thermometer at hand, we report how the fluid temperature changes as a function of time while following the fluid, i.e.,

$$\left(\frac{dT}{dt} \right)_{\mathbf{R}} = \frac{DT}{Dt}, \tag{1.1-4}$$

  where $\mathbf{R}$ is the *material position vector* locating the fluid particle at $t = 0$. It represents the coordinates used to identify a given particle.

The relationship between the partial and material time derivatives is expressed as

$$\frac{D\varphi}{Dt} = \frac{\partial \varphi}{\partial t} + \mathbf{v} \cdot \nabla \varphi. \tag{1.1-5}$$

The time rate of change of velocity of a fluid particle is the acceleration **a**, given by

$$\mathbf{a} = \left(\frac{d\mathbf{v}}{dt}\right)_{\mathbf{R}} = \frac{D\mathbf{v}}{Dt} = \underbrace{\frac{\partial \mathbf{v}}{\partial t}}_{\substack{\text{Local} \\ \text{acceleration}}} + \underbrace{\mathbf{v} \cdot \nabla \mathbf{v}}_{\substack{\text{Convective} \\ \text{acceleration}}}. \tag{1.1-6}$$

The first term on the right-hand side of Eq. (1.1-6) is the rate of change of velocity at a fixed point in space and is called the *local acceleration*. The second term is called the *convective acceleration* and it depends upon both the magnitude of the velocity and the velocity gradients. When the flow is steady, the local acceleration is zero. However, the value of the convective acceleration is dependent on the flow geometry. For example, for a steady flow through a converging nozzle, fluid velocity increases in the flow direction as a result of the decrease in the cross-sectional area, leading to a nonzero value for the convective acceleration.

In the cylindrical coordinate system shown in Figure A.3 in Appendix A, Eq. (1.1-6) is expressed as

$$\mathbf{a} = \left(\frac{Dv_r}{Dt} - \frac{v_\theta^2}{r}\right)\mathbf{e}_r + \left(\frac{Dv_\theta}{Dt} + \frac{v_r v_\theta}{r}\right)\mathbf{e}_\theta + \frac{Dv_z}{Dt}\mathbf{e}_z. \tag{1.1-7}$$

The term "$-v_\theta^2/r$" is the contribution due to *centrifugal force*; it is an effective force in the r-direction. The term "$v_r v_\theta/r$" is sometimes called a *Coriolis acceleration*; it produces an effective force in the θ-direction.

In the spherical coordinate system shown in Figure A.4 in Appendix A, Eq. (1.1-6) is expressed as

$$\mathbf{a} = \left(\frac{Dv_r}{Dt} - \frac{v_\theta^2 + v_\phi^2}{r}\right)\mathbf{e}_r + \left(\frac{Dv_\theta}{Dt} + \frac{v_r v_\theta}{r} - \frac{v_\phi^2 \cot \theta}{r}\right)\mathbf{e}_\theta$$

$$+ \left(\frac{Dv_\phi}{Dt} + \frac{v_\phi v_r}{r} + \frac{v_\theta v_\phi \cot \theta}{r}\right)\mathbf{e}_\phi. \tag{1.1-8}$$

1.2   Governing Equations

1.2.1   *Equation of continuity*

The equation of continuity is nothing but the conservation of mass, expressed as

$$\frac{\partial \rho}{\partial t} + \nabla \cdot \rho \mathbf{v} = 0. \tag{1.2-1}$$

The equation of continuity in various coordinate systems is shown in Table 1.1. With the help of Eq. (1.1-5), Eq. (1.2-1) takes the following form:

$$\frac{D\rho}{Dt} + \rho \left(\nabla \cdot \mathbf{v} \right) = 0. \tag{1.2-2}$$

Rearrangement of Eq. (1.2-2) and noting that $\rho = 1/\widehat{V}$ give

$$\nabla \cdot \mathbf{v} = -\frac{1}{\rho} \frac{D\rho}{Dt} = \frac{1}{V} \frac{DV}{Dt}. \tag{1.2-3}$$

Therefore, $\nabla \cdot \mathbf{v}$ can be interpreted as the rate of expansion of a material volume per unit volume, known as the *dilatation rate*.

For a fluid of constant density, i.e., an incompressible fluid,[1] Eq. (1.2-1) simplifies to

$$\nabla \cdot \mathbf{v} = 0. \tag{1.2-4}$$

A flow satisfying Eq. (1.2-4) is called an *incompressible flow*. Note that compressible fluids may also satisfy Eq. (1.2-4) depending on the flow geometry and conditions.

Example 1   Express the term Dv_r/Dt in Eq. (1.1-7).

Solution   Application of Eq. (1.1-5) gives

$$\frac{Dv_r}{Dt} = \frac{\partial v_r}{\partial t} + \mathbf{v} \cdot \nabla v_r.$$

The gradient of a scalar field in cylindrical coordinate system is given by Eq. (A.8-2) in Appendix A. Hence,

$$\frac{Dv_r}{Dt} = \frac{\partial v_r}{\partial t} + (v_r \mathbf{e}_r + v_\theta \mathbf{e}_\theta + v_z \mathbf{e}_z) \cdot \left(\frac{\partial v_r}{\partial r} \mathbf{e}_r + \frac{1}{r} \frac{\partial v_r}{\partial \theta} \mathbf{e}_\theta + \frac{\partial v_r}{\partial z} \mathbf{e}_z \right)$$

$$= \frac{\partial v_r}{\partial t} + v_r \frac{\partial v_r}{\partial r} + \frac{v_\theta}{r} \frac{\partial v_r}{\partial \theta} + v_z \frac{\partial v_r}{\partial z}.$$

---

[1]There is no such thing as an incompressible fluid. It is an approximation used in engineering calculations.

Table 1.1   The equation of continuity in various coordinate systems.

Cartesian coordinates

$$\frac{\partial \rho}{\partial t} + \frac{\partial}{\partial x}(\rho v_x) + \frac{\partial}{\partial y}(\rho v_y) + \frac{\partial}{\partial z}(\rho v_z) = 0 \tag{A}$$

Cylindrical coordinates

$$\frac{\partial \rho}{\partial t} + \frac{1}{r}\frac{\partial}{\partial r}(\rho r v_r) + \frac{1}{r}\frac{\partial}{\partial \theta}(\rho v_\theta) + \frac{\partial}{\partial z}(\rho v_z) = 0 \tag{B}$$

Spherical coordinates

$$\frac{\partial \rho}{\partial t} + \frac{1}{r^2}\frac{\partial}{\partial r}(\rho r^2 v_r) + \frac{1}{r\sin\theta}\frac{\partial}{\partial \theta}(\rho v_\theta \sin\theta) + \frac{1}{r\sin\theta}\frac{\partial}{\partial \phi}(\rho v_\phi) = 0 \tag{C}$$

Example 2   The velocity vector in a flow field is expressed as

$$\mathbf{v} = 2xy\,\mathbf{e}_x - yt\,\mathbf{e}_y + 3z\,\mathbf{e}_z.$$

(a) Is this flow compressible or incompressible?
(b) Express the components of the fluid acceleration **a**.

Solution

(a) Equation (1.2-4) states that $\nabla \cdot \mathbf{v} = 0$ for an incompressible flow. The divergence of a vector field in the Cartesian coordinate system is given by Eq. (A.5-3) in Appendix A. Hence,

$$\nabla \cdot \mathbf{v} = \frac{\partial v_x}{\partial x} + \frac{\partial v_y}{\partial y} + \frac{\partial v_z}{\partial z} = 2y - t + 3 \neq 0 \Rightarrow \text{compressible flow.}$$

(b) In the Cartesian coordinate system, Eq. (1.1-6) is expressed as

$$\mathbf{a} = \frac{\partial v_j}{\partial t}\mathbf{e}_j + v_i\mathbf{e}_i \cdot \frac{\partial v_j}{\partial x_k}\mathbf{e}_k\mathbf{e}_j = \frac{\partial v_j}{\partial t}\mathbf{e}_j + v_i\frac{\partial v_j}{\partial x_k}\underbrace{(\mathbf{e}_i \cdot \mathbf{e}_k)}_{\delta_{ik}}\mathbf{e}_j$$

$$= \left(\frac{\partial v_j}{\partial t} + v_i\frac{\partial v_j}{\partial x_i}\right)\mathbf{e}_j.$$

Therefore,

$$a_x = \frac{\partial v_x}{\partial t} + v_x\frac{\partial v_x}{\partial x} + v_y\frac{\partial v_x}{\partial y} + v_z\frac{\partial v_x}{\partial z} = 4xy^2 - 2xyt,$$

$$a_y = \frac{\partial v_y}{\partial t} + v_x\frac{\partial v_y}{\partial x} + v_y\frac{\partial v_y}{\partial y} + v_z\frac{\partial v_y}{\partial z} = -y + yt^2,$$

$$a_z = \frac{\partial v_z}{\partial t} + v_x\frac{\partial v_z}{\partial x} + v_y\frac{\partial v_z}{\partial y} + v_z\frac{\partial v_z}{\partial z} = 9z.$$

1.2.2 Equation of motion

The equation of motion is nothing but Newton's second law of motion, i.e.,

$$(\text{Mass})(\text{Acceleration}) = \sum \text{Forces acting on a body}. \qquad (1.2\text{-}5)$$

Forces can be classified as *body forces* that act at a distance, and *surface forces* that act by direct contact. Gravitational force, electrostatic force, and electromagnetic force are examples of body forces. Surface forces, on the other hand, are classified as normal forces (pressure) and tangential forces (shear stress). In vector notation, the equation of motion is written as

$$\rho \frac{D\mathbf{v}}{Dt} = -\nabla \cdot \boldsymbol{\pi} + \rho \mathbf{f}, \qquad (1.2\text{-}6)$$

where $\boldsymbol{\pi}$ is the stress tensor[2] and $\mathbf{f}$ is the body force per unit mass. Note that Eq. (1.2-6) is expressed as

$$\left(\frac{\text{Mass}}{\text{Volume}}\right)(\text{Acceleration}) = \left(\frac{\text{Surface forces}}{\text{Volume}}\right) + \left(\frac{\text{Body forces}}{\text{Volume}}\right).$$
$$(1.2\text{-}7)$$

While the equation of continuity, Eq. (1.2-1), is a scalar equation, the equation of motion, Eq. (1.2-6), is a vector equation. Hence, it has three components.

When the fluid is at rest, the stress is hydrostatic and is determined by a scalar quantity P_t, known as the *thermodynamic pressure*. For single-phase and single-component systems, the Gibbs phase rule indicates that the state of a system can be determined by specifying two independent intensive variables, such as density, ρ, and temperature, T. Thus, the thermodynamic pressure is specified by an equation of state of the form

$$P_t = P_t(\rho, T). \qquad (1.2\text{-}8)$$

The thermodynamic pressure always acts normal to the surface and is independent of orientation. Thus, it is possible to express the stress tensor as

$$\boldsymbol{\pi} = P_t \mathbf{I} + \boldsymbol{\tau}, \qquad (1.2\text{-}9)$$

where $\mathbf{I}$ is an identity tensor and $\boldsymbol{\tau}$ is the *shear* (or *viscous*) *stress tensor*. Note that $\boldsymbol{\tau}$ vanishes when there is no motion, i.e.,

$$\boldsymbol{\tau} = 0 \quad \text{For fluids at rest.} \qquad (1.2\text{-}10)$$

---

[2]The stress tensor is symmetric, i.e., $\boldsymbol{\pi} = \boldsymbol{\pi}^{\mathrm{T}}$.

The normal stress is the sum of the thermodynamic pressure and the normal shear stress, i.e.,

$$\pi_{11} = P_t + \tau_{11}, \tag{1.2-11}$$

$$\pi_{22} = P_t + \tau_{22}, \tag{1.2-12}$$

$$\pi_{33} = P_t + \tau_{33}. \tag{1.2-13}$$

The normal stress can take different values for different directions. The average of the normal stress is known as the *mechanical pressure*, P_m, i.e.,

$$P_m = \frac{\pi_{11} + \pi_{22} + \pi_{33}}{3}. \tag{1.2-14}$$

Thus, substitution of Eqs. (1.2-11)–(1.2-13) into Eq. (1.2-14) leads to

$$P_m = P_t + \frac{\tau_{11} + \tau_{22} + \tau_{33}}{3}. \tag{1.2-15}$$

When a fluid is moving, the difference between the mechanical and thermodynamic pressures is dependent on the fluid type as follows:

- For an incompressible fluid, density is constant. Thus, an incompressible fluid does not have a thermodynamic pressure and the pressure variable must be interpreted as the mechanical pressure.
- For a compressible fluid, the difference between the thermodynamic and mechanical pressures is assumed to be a linear function of the dilatation rate (or rate of expansion), i.e.,

$$P_t - P_m = \kappa \left(\nabla \cdot \mathbf{v} \right), \tag{1.2-16}$$

where κ is called the *bulk* (or *dilatational*) *viscosity*.

For most fluids κ is considered zero.[3] In this case, Eq. (1.2-16) implies that $P_t = P_m$, i.e., there is no need to distinguish between the mechanical and thermodynamic pressures, and Eq. (1.2-15) implies that the summation of the normal shear stresses, i.e., $\tau_{11} + \tau_{22} + \tau_{33}$, is zero.

Taking $P_t = P_m = P$ and substituting Eq. (1.2-9) into Eq. (1.2-6) yield

$$\rho \frac{D\mathbf{v}}{Dt} = -\nabla P - \nabla \cdot \boldsymbol{\tau} + \rho \mathbf{g}. \tag{1.2-17}$$

In writing Eq. (1.2-17) it is implicitly assumed that the gravitational force is the only body force acting on the system, i.e., $\mathbf{f} = \mathbf{g}$. Equation (1.2-17) holds for any fluid and for any coordinate system.

---

[3]This is known as the *Stokes assumption*.

Table 1.2   Components of $\boldsymbol{\tau}$ for Newtonian fluids in Cartesian coordinates (x, y, z).

$$\tau_{xx} = -\mu \left[2 \frac{\partial v_x}{\partial x} - \frac{2}{3} (\nabla \cdot \mathbf{v}) \right] \qquad \text{(A)}$$

$$\tau_{yy} = -\mu \left[2 \frac{\partial v_y}{\partial y} - \frac{2}{3} (\nabla \cdot \mathbf{v}) \right] \qquad \text{(B)}$$

$$\tau_{zz} = -\mu \left[2 \frac{\partial v_z}{\partial z} - \frac{2}{3} (\nabla \cdot \mathbf{v}) \right] \qquad \text{(C)}$$

$$\tau_{xy} = \tau_{yx} = -\mu \left(\frac{\partial v_x}{\partial y} + \frac{\partial v_y}{\partial x} \right) \qquad \text{(D)}$$

$$\tau_{yz} = \tau_{zy} = -\mu \left(\frac{\partial v_y}{\partial z} + \frac{\partial v_z}{\partial y} \right) \qquad \text{(E)}$$

$$\tau_{zx} = \tau_{xz} = -\mu \left(\frac{\partial v_z}{\partial x} + \frac{\partial v_x}{\partial z} \right) \qquad \text{(F)}$$

$$\nabla \cdot \mathbf{v} = \frac{\partial v_x}{\partial x} + \frac{\partial v_y}{\partial y} + \frac{\partial v_z}{\partial z} \qquad \text{(G)}$$

If the viscous effects are negligible, i.e., $\nabla \cdot \boldsymbol{\tau} \simeq 0$, Eq. (1.2-17) reduces to

$$\rho \frac{D\mathbf{v}}{Dt} = -\nabla P + \rho \mathbf{g}, \qquad (1.2\text{-}18)$$

which is known as the *Euler equation*.

For a Newtonian fluid, a constitutive equation[4] for $\boldsymbol{\tau}$ is expressed as

$$\boldsymbol{\tau} = -\mu \dot{\boldsymbol{\gamma}} + \left(\frac{2}{3} \mu - \kappa \right) (\nabla \cdot \mathbf{v}) \, \mathbf{I}, \qquad (1.2\text{-}19)$$

where μ is the *viscosity* and $\dot{\boldsymbol{\gamma}}$ is called the *rate of deformation tensor*, defined by

$$\dot{\boldsymbol{\gamma}} = \nabla \mathbf{v} + (\nabla \mathbf{v})^{\mathrm{T}}. \qquad (1.2\text{-}20)$$

Taking $\kappa = 0$, the components of the shear stress tensor for Newtonian fluids in Cartesian, cylindrical, and spherical coordinate systems are given in Tables 1.2–1.4, respectively.

---

[4] Any mathematical description of the response of a material to spatial gradients is called a constitutive equation. The coefficients appearing in constitutive equations are obtained from experiments.

Table 1.3   Components of $\boldsymbol{\tau}$ for Newtonian fluids in cylindrical coordinates (r, θ, z).

$$\tau_{rr} = -\mu \left[2 \frac{\partial v_r}{\partial r} - \frac{2}{3} (\nabla \cdot \mathbf{v}) \right] \tag{A}$$

$$\tau_{\theta\theta} = -\mu \left[2 \left(\frac{1}{r} \frac{\partial v_\theta}{\partial \theta} + \frac{v_r}{r} \right) - \frac{2}{3} (\nabla \cdot \mathbf{v}) \right] \tag{B}$$

$$\tau_{zz} = -\mu \left[2 \frac{\partial v_z}{\partial z} - \frac{2}{3} (\nabla \cdot \mathbf{v}) \right] \tag{C}$$

$$\tau_{r\theta} = \tau_{\theta r} = -\mu \left[r \frac{\partial}{\partial r} \left(\frac{v_\theta}{r} \right) + \frac{1}{r} \frac{\partial v_r}{\partial \theta} \right] \tag{D}$$

$$\tau_{\theta z} = \tau_{z\theta} = -\mu \left(\frac{\partial v_\theta}{\partial z} + \frac{1}{r} \frac{\partial v_z}{\partial \theta} \right) \tag{E}$$

$$\tau_{zr} = \tau_{rz} = -\mu \left(\frac{\partial v_z}{\partial r} + \frac{\partial v_r}{\partial z} \right) \tag{F}$$

$$\nabla \cdot \mathbf{v} = \frac{1}{r} \frac{\partial}{\partial r} (r v_r) + \frac{1}{r} \frac{\partial v_\theta}{\partial \theta} + \frac{\partial v_z}{\partial z} \tag{G}$$

Table 1.4   Components of $\boldsymbol{\tau}$ for Newtonian fluids in spherical coordinates (r, θ, ϕ).

$$\tau_{rr} = -\mu \left[2 \frac{\partial v_r}{\partial r} - \frac{2}{3} (\nabla \cdot \mathbf{v}) \right] \tag{A}$$

$$\tau_{r\theta} = -\mu \left[2 \left(\frac{1}{r} \frac{\partial v_\theta}{\partial \theta} + \frac{v_r}{r} \right) - \frac{2}{3} (\nabla \cdot \mathbf{v}) \right] \tag{B}$$

$$\tau_{\phi\phi} = -\mu \left[2 \left(\frac{1}{r \sin\theta} \frac{\partial v_\phi}{\partial \phi} + \frac{v_r}{r} + \frac{v_\theta \cot\theta}{r} \right) - \frac{2}{3} (\nabla \cdot \mathbf{v}) \right] \tag{C}$$

$$\tau_{r\theta} = \tau_{\theta r} = -\mu \left[r \frac{\partial}{\partial r} \left(\frac{v_\theta}{r} \right) + \frac{1}{r} \frac{\partial v_r}{\partial \theta} \right] \tag{D}$$

$$\tau_{\theta\phi} = \tau_{\phi\theta} = -\mu \left[\frac{\sin\theta}{r} \frac{\partial}{\partial \theta} \left(\frac{v_\phi}{\sin\theta} \right) + \frac{1}{r \sin\theta} \frac{\partial v_\theta}{\partial \phi} \right] \tag{E}$$

$$\tau_{\phi r} = \tau_{r\phi} = -\mu \left[\frac{1}{r \sin\theta} \frac{\partial v_r}{\partial \phi} + r \frac{\partial}{\partial r} \left(\frac{v_\phi}{r} \right) \right] \tag{F}$$

$$\nabla \cdot \mathbf{v} = \frac{1}{r^2} \frac{\partial}{\partial r} (r^2 v_r) + \frac{1}{r \sin\theta} \frac{\partial}{\partial \theta} (v_\theta \sin\theta) + \frac{1}{r \sin\theta} \frac{\partial v_\phi}{\partial \phi} \tag{G}$$

For an incompressible fluid, Eq. (1.2-19) reduces to

$$\boldsymbol{\tau} = -\mu\,\dot{\boldsymbol{\gamma}} = -\mu\left[\nabla\mathbf{v} + (\nabla\mathbf{v})^{\mathrm{T}}\right]. \tag{1.2-21}$$

Substitution of Eq. (1.2-21) into Eq. (1.2-17) leads to

$$\rho\frac{D\mathbf{v}}{Dt} = -\nabla P + \mu\nabla^2\mathbf{v} + \rho\mathbf{g}, \tag{1.2-22}$$

which is also called the *Navier–Stokes equation* in the literature. Keep in mind that both density and viscosity are considered constant in the derivation of Eq. (1.2-22).

The components of Eqs. (1.2-17) and (1.2-22) in Cartesian, cylindrical, and spherical coordinate systems are given in Tables 1.5–1.7, respectively.

Sometimes it is convenient to combine pressure and gravitational terms as follows. Since $\mathbf{g}$ is a conservative force,[5] i.e., $\nabla\times\mathbf{g} = 0$, it can be expressed as

$$\mathbf{g} = -\nabla\widehat{\phi}, \tag{1.2-23}$$

where $\widehat{\phi}$ is the gravitational potential energy per unit mass. The minus sign in Eq. (1.2-23) indicates that the positive z-direction is in the direction opposite to gravity. Substitution of Eq. (1.2-23) into Eq. (1.2-22) and rearrangement lead to

$$\rho\frac{D\mathbf{v}}{Dt} = -\nabla\mathcal{P} + \mu\nabla^2\mathbf{v}. \tag{1.2-24}$$

The term $\mathcal{P}$ in Eq. (1.2-24) is called *modified pressure*,[6] defined by

$$\mathcal{P} = P + \rho\widehat{\phi} \tag{1.2-25}$$

Physical interpretation of modified pressure will be explained in Section 3.5.

1.2.3   *Mechanical energy equation*

Power requirement is equal to the rate at which work is done to force the fluid through the system, i.e.,

$$\text{Power} = \dot{W} = \frac{\text{Work}}{\text{Time}} = \frac{(\text{Force})(\text{Distance})}{\text{Time}}$$

$$= (\text{Force})(\text{Velocity}) = \mathbf{F}\cdot\mathbf{v}, \tag{1.2-26}$$

---

[5] A force $\mathbf{F}$ is called conservative if the work done by the force is a state function. In mathematical terms, $\nabla\times\mathbf{F} = 0$.

[6] In the literature, $\mathcal{P}$ is also called equivalent pressure, dynamic pressure, and piezometric pressure.

Table 1.5 The equation of motion in Cartesian coordinates (x, y, z).

In terms of shear stress components:

x-component

$$\rho \left(\frac{\partial v_x}{\partial t} + v_x \frac{\partial v_x}{\partial x} + v_y \frac{\partial v_x}{\partial y} + v_z \frac{\partial v_x}{\partial z} \right)$$

$$= -\frac{\partial P}{\partial x} - \left(\frac{\partial \tau_{xx}}{\partial x} + \frac{\partial \tau_{yx}}{\partial y} + \frac{\partial \tau_{zx}}{\partial z} \right) + \rho g_x \qquad \text{(A)}$$

y-component

$$\rho \left(\frac{\partial v_y}{\partial t} + v_x \frac{\partial v_y}{\partial x} + v_y \frac{\partial v_y}{\partial y} + v_z \frac{\partial v_y}{\partial z} \right)$$

$$= -\frac{\partial P}{\partial y} - \left(\frac{\partial \tau_{xy}}{\partial x} + \frac{\partial \tau_{yy}}{\partial y} + \frac{\partial \tau_{zy}}{\partial z} \right) + \rho g_y \qquad \text{(B)}$$

z-component

$$\rho \left(\frac{\partial v_z}{\partial t} + v_x \frac{\partial v_z}{\partial x} + v_y \frac{\partial v_z}{\partial y} + v_z \frac{\partial v_z}{\partial z} \right)$$

$$= -\frac{\partial P}{\partial z} - \left(\frac{\partial \tau_{xz}}{\partial x} + \frac{\partial \tau_{yz}}{\partial y} + \frac{\partial \tau_{zz}}{\partial z} \right) + \rho g_z \qquad \text{(C)}$$

For a Newtonian fluid with constant ρ and μ:

x-component

$$\rho \left(\frac{\partial v_x}{\partial t} + v_x \frac{\partial v_x}{\partial x} + v_y \frac{\partial v_x}{\partial y} + v_z \frac{\partial v_x}{\partial z} \right)$$

$$= -\frac{\partial P}{\partial x} + \mu \left(\frac{\partial^2 v_x}{\partial x^2} + \frac{\partial^2 v_x}{\partial y^2} + \frac{\partial^2 v_x}{\partial z^2} \right) + \rho g_x \qquad \text{(D)}$$

y-component

$$\rho \left(\frac{\partial v_y}{\partial t} + v_x \frac{\partial v_y}{\partial x} + v_y \frac{\partial v_y}{\partial y} + v_z \frac{\partial v_y}{\partial z} \right)$$

$$= -\frac{\partial P}{\partial y} + \mu \left(\frac{\partial^2 v_y}{\partial x^2} + \frac{\partial^2 v_y}{\partial y^2} + \frac{\partial^2 v_y}{\partial z^2} \right) + \rho g_y \qquad \text{(E)}$$

z-component

$$\rho \left(\frac{\partial v_z}{\partial t} + v_x \frac{\partial v_z}{\partial x} + v_y \frac{\partial v_z}{\partial y} + v_z \frac{\partial v_z}{\partial z} \right)$$

$$= -\frac{\partial P}{\partial z} + \mu \left(\frac{\partial^2 v_z}{\partial x^2} + \frac{\partial^2 v_z}{\partial y^2} + \frac{\partial^2 v_z}{\partial z^2} \right) + \rho g_z \qquad \text{(F)}$$

indicating that the rate of work is the projection of the force along the direction of the velocity. Only this component of force increases the kinetic energy of the particle. Therefore, the dot product of the equation of motion with velocity results in the *mechanical* (or *kinetic*) *energy equation*, i.e.,

$$\text{Mechanical energy equation} = \mathbf{v} \cdot (\text{Equation of motion}). \qquad (1.2\text{-}27)$$

Note that the mechanical energy equation is not an independent equation. Thus, the dot product of Eq. (1.2-6) with velocity yields

$$\rho \mathbf{v} \cdot \frac{D\mathbf{v}}{Dt} = -\mathbf{v} \cdot (\nabla \cdot \boldsymbol{\pi}) + \rho \left(\mathbf{v} \cdot \mathbf{f} \right). \qquad (1.2\text{-}28)$$

Table 1.6   The equation of motion in cylindrical coordinates (r, θ, z).

In terms of shear stress components:

r-component $\rho \left(\dfrac{\partial v_r}{\partial t} + v_r \dfrac{\partial v_r}{\partial r} + \dfrac{v_\theta}{r} \dfrac{\partial v_r}{\partial \theta} - \dfrac{v_\theta^2}{r} + v_z \dfrac{\partial v_r}{\partial z} \right)$

$$= -\frac{\partial P}{\partial r} - \left[\frac{1}{r} \frac{\partial}{\partial r}(r\tau_{rr}) + \frac{1}{r} \frac{\partial \tau_{r\theta}}{\partial \theta} - \frac{\tau_{\theta\theta}}{r} + \frac{\partial \tau_{rz}}{\partial z} \right] + \rho g_r \qquad (A)$$

θ-component $\rho \left(\dfrac{\partial v_\theta}{\partial t} + v_r \dfrac{\partial v_\theta}{\partial r} + \dfrac{v_\theta}{r} \dfrac{\partial v_\theta}{\partial \theta} + \dfrac{v_r v_\theta}{r} + v_z \dfrac{\partial v_\theta}{\partial z} \right)$

$$= -\frac{1}{r}\frac{\partial P}{\partial \theta} - \left[\frac{1}{r^2} \frac{\partial}{\partial r}(r^2 \tau_{r\theta}) + \frac{1}{r} \frac{\partial \tau_{\theta\theta}}{\partial \theta} + \frac{\partial \tau_{\theta z}}{\partial z} \right] + \rho g_\theta \qquad (B)$$

z-component $\rho \left(\dfrac{\partial v_z}{\partial t} + v_r \dfrac{\partial v_z}{\partial r} + \dfrac{v_\theta}{r} \dfrac{\partial v_z}{\partial \theta} + v_z \dfrac{\partial v_z}{\partial z} \right)$

$$= -\frac{\partial P}{\partial z} - \left[\frac{1}{r} \frac{\partial}{\partial r}(r\tau_{rz}) + \frac{1}{r} \frac{\partial \tau_{\theta z}}{\partial \theta} + \frac{\partial \tau_{zz}}{\partial z} \right] + \rho g_z \qquad (C)$$

For a Newtonian fluid with constant ρ and μ:

r-component $\rho \left(\dfrac{\partial v_r}{\partial t} + v_r \dfrac{\partial v_r}{\partial r} + \dfrac{v_\theta}{r} \dfrac{\partial v_r}{\partial \theta} - \dfrac{v_\theta^2}{r} + v_z \dfrac{\partial v_r}{\partial z} \right)$

$$= -\frac{\partial P}{\partial r} + \mu \left\{ \frac{\partial}{\partial r} \left[\frac{1}{r} \frac{\partial}{\partial r}(r v_r) \right] + \frac{1}{r^2} \frac{\partial^2 v_r}{\partial \theta^2} - \frac{2}{r^2} \frac{\partial v_\theta}{\partial \theta} + \frac{\partial^2 v_r}{\partial z^2} \right\}$$
$$+ \rho g_r \qquad (D)$$

θ-component $\rho \left(\dfrac{\partial v_\theta}{\partial t} + v_r \dfrac{\partial v_\theta}{\partial r} + \dfrac{v_\theta}{r} \dfrac{\partial v_\theta}{\partial \theta} + \dfrac{v_r v_\theta}{r} + v_z \dfrac{\partial v_\theta}{\partial z} \right)$

$$= -\frac{1}{r}\frac{\partial P}{\partial \theta} + \mu \left\{ \frac{\partial}{\partial r} \left[\frac{1}{r} \frac{\partial}{\partial r}(r v_\theta) \right] + \frac{1}{r^2} \frac{\partial^2 v_\theta}{\partial \theta^2} + \frac{2}{r^2} \frac{\partial v_r}{\partial \theta} + \frac{\partial^2 v_\theta}{\partial z^2} \right\}$$
$$+ \rho g_\theta \qquad (E)$$

z-component $\rho \left(\dfrac{\partial v_z}{\partial t} + v_r \dfrac{\partial v_z}{\partial r} + \dfrac{v_\theta}{r} \dfrac{\partial v_z}{\partial \theta} + v_z \dfrac{\partial v_z}{\partial z} \right)$

$$= -\frac{\partial P}{\partial z} + \mu \left[\frac{1}{r} \frac{\partial}{\partial r} \left(r \frac{\partial v_z}{\partial r} \right) + \frac{1}{r^2} \frac{\partial^2 v_z}{\partial \theta^2} + \frac{\partial^2 v_z}{\partial z^2} \right] + \rho g_z \qquad (F)$$

Noting that

$$\frac{D(\mathbf{v} \cdot \mathbf{v})}{Dt} = \mathbf{v} \cdot \frac{D\mathbf{v}}{Dt} + \frac{D\mathbf{v}}{Dt} \cdot \mathbf{v} \Rightarrow \mathbf{v} \cdot \frac{D\mathbf{v}}{Dt} = \frac{D(\frac{1}{2}v^2)}{Dt}, \qquad (1.2\text{-}29)$$

Eq. (1.2-28) takes the form

$$\rho \frac{D(\frac{1}{2}v^2)}{Dt} = -\mathbf{v} \cdot (\nabla \cdot \boldsymbol{\pi}) + \rho (\mathbf{v} \cdot \mathbf{f}). \qquad (1.2\text{-}30)$$

Table 1.7   The equation of motion in spherical coordinates (r, θ, ϕ).

In terms of shear stress components:

r-component $\rho \left(\dfrac{\partial v_r}{\partial t} + v_r \dfrac{\partial v_r}{\partial r} + \dfrac{v_\theta}{r} \dfrac{\partial v_r}{\partial \theta} + \dfrac{v_\phi}{r \sin \theta} \dfrac{\partial v_r}{\partial \phi} - \dfrac{v_\theta^2 + v_\phi^2}{r} \right)$

$$= -\frac{\partial P}{\partial r} - \left[\frac{1}{r^2} \frac{\partial}{\partial r} (r^2 \tau_{rr}) + \frac{1}{r \sin \theta} \frac{\partial}{\partial \theta} (\tau_{r\theta} \sin \theta) \right.$$

$$\left. + \frac{1}{r \sin \theta} \frac{\partial \tau_{r\phi}}{\partial \phi} - \frac{\tau_{\theta\theta} + \tau_{\phi\phi}}{r} \right] + \rho g_r \qquad (A)$$

θ-component $\rho \left(\dfrac{\partial v_\theta}{\partial t} + v_r \dfrac{\partial v_\theta}{\partial r} + \dfrac{v_\theta}{r} \dfrac{\partial v_\theta}{\partial \theta} + \dfrac{v_\phi}{r \sin \theta} \dfrac{\partial v_\theta}{\partial \phi} + \dfrac{v_r v_\theta}{r} - \dfrac{v_\phi^2 \cot \theta}{r} \right)$

$$= -\frac{1}{r} \frac{\partial P}{\partial \theta} - \left[\frac{1}{r^2} \frac{\partial}{\partial r} (r^2 \tau_{r\theta}) + \frac{1}{r \sin \theta} \frac{\partial}{\partial \theta} (\tau_{\theta\theta} \sin \theta) \right.$$

$$\left. + \frac{1}{r \sin \theta} \frac{\partial \tau_{\theta\phi}}{\partial \phi} + \frac{\tau_{r\theta}}{r} - \frac{\cot \theta}{r} \tau_{\phi\phi} \right] + \rho g_\theta \qquad (B)$$

ϕ-component $\rho \left(\dfrac{\partial v_\phi}{\partial t} + v_r \dfrac{\partial v_\phi}{\partial r} + \dfrac{v_\theta}{r} \dfrac{\partial v_\phi}{\partial \theta} + \dfrac{v_\phi}{r \sin \theta} \dfrac{\partial v_\phi}{\partial \phi} + \dfrac{v_\phi v_r}{r} + \dfrac{v_\theta v_\phi}{r} \cot \theta \right)$

$$= -\frac{1}{r \sin \theta} \frac{\partial P}{\partial \phi} - \left[\frac{1}{r^2} \frac{\partial}{\partial r} (r^2 \tau_{r\phi}) + \frac{1}{r} \frac{\partial \tau_{\theta\phi}}{\partial \theta} + \frac{1}{r \sin \theta} \frac{\partial \tau_{\phi\phi}}{\partial \phi} \right.$$

$$\left. + \frac{\tau_{r\phi}}{r} + \frac{2 \cot \theta}{r} \tau_{\theta\phi} \right] + \rho g_\phi \qquad (C)$$

For a Newtonian fluid with constant ρ and μ[a]:

r-component $\rho \left(\dfrac{\partial v_r}{\partial t} + v_r \dfrac{\partial v_r}{\partial r} + \dfrac{v_\theta}{r} \dfrac{\partial v_r}{\partial \theta} + \dfrac{v_\phi}{r \sin \theta} \dfrac{\partial v_r}{\partial \phi} - \dfrac{v_\theta^2 + v_\phi^2}{r} \right)$

$$= -\frac{\partial P}{\partial r} + \mu \left(\nabla^2 v_r - \frac{2}{r^2} v_r - \frac{2}{r^2} \frac{\partial v_\theta}{\partial \theta} - \frac{2}{r^2} v_\theta \cot \theta \right.$$

$$\left. - \frac{2}{r^2 \sin \theta} \frac{\partial v_\phi}{\partial \phi} \right) + \rho g_r \qquad (D)$$

θ-component $\rho \left(\dfrac{\partial v_\theta}{\partial t} + v_r \dfrac{\partial v_\theta}{\partial r} + \dfrac{v_\theta}{r} \dfrac{\partial v_\theta}{\partial \theta} + \dfrac{v_\phi}{r \sin \theta} \dfrac{\partial v_\theta}{\partial \phi} + \dfrac{v_r v_\theta}{r} - \dfrac{v_\phi^2 \cot \theta}{r} \right)$

$$= -\frac{1}{r} \frac{\partial P}{\partial \theta} + \mu \left(\nabla^2 v_\theta + \frac{2}{r^2} \frac{\partial v_r}{\partial \theta} - \frac{v_\theta}{r^2 \sin^2 \theta} \right.$$

$$\left. - \frac{2 \cos \theta}{r^2 \sin^2 \theta} \frac{\partial v_\phi}{\partial \phi} \right) + \rho g_\theta \qquad (E)$$

(Continued)

Table 1.7   (*Continued*)

ϕ-component $\rho \left(\dfrac{\partial v_\phi}{\partial t} + v_r \dfrac{\partial v_\phi}{\partial r} + \dfrac{v_\theta}{r} \dfrac{\partial v_\phi}{\partial \theta} + \dfrac{v_\phi}{r \sin \theta} \dfrac{\partial v_\phi}{\partial \phi} + \dfrac{v_\phi v_r}{r} + \dfrac{v_\phi v_\phi}{r} \cot \theta \right)$

$$= -\dfrac{1}{r \sin \theta} \dfrac{\partial P}{\partial \phi} + \mu \left(\nabla^2 v_\phi - \dfrac{v_\phi}{r^2 \sin^2 \theta} + \dfrac{2}{r^2 \sin \theta} \dfrac{\partial v_r}{\partial \phi} \right.$$

$$\left. + \dfrac{2 \cos \theta}{r^2 \sin^2 \theta} \dfrac{\partial v_\theta}{\partial \phi} \right) + \rho g_\phi \qquad \text{(F)}$$

---

[a]The Laplacian operator in Eqs. (D), (E), and (F) is given by

$$\nabla^2 = \dfrac{1}{r^2} \dfrac{\partial}{\partial r} \left(r^2 \dfrac{\partial}{\partial r} \right) + \dfrac{1}{r^2 \sin \theta} \dfrac{\partial}{\partial \theta} \left(\sin \theta \dfrac{\partial}{\partial \theta} \right) + \dfrac{1}{r^2 \sin^2 \theta} \dfrac{\partial^2}{\partial \phi^2}.$$

Taking $P_t = P$ and substituting Eq. (1.2-9) into Eq. (1.2-30) give

$$\rho \dfrac{D(\frac{1}{2} v^2)}{Dt} = -\mathbf{v} \cdot \nabla P - \mathbf{v} \cdot (\nabla \cdot \boldsymbol{\tau}) + \rho \left(\mathbf{v} \cdot \mathbf{f} \right). \qquad (1.2\text{-}31)$$

The first, second, and third terms on the right-hand side of Eq. (1.2-31) represent the rate of work done by pressure force per unit volume, the rate of work done by viscous force per unit volume, and the rate of work done by body forces per unit volume, respectively. Thus, all of the work of the surface and body forces goes to accelerate the fluid and increase its kinetic energy.

1.3   Boundary Conditions

Equations of continuity and motion constitute the governing differential equations (ordinary or partial) for the velocity and pressure fields. These equations must be solved with appropriate initial and boundary conditions.

1.3.1   *Conditions at a solid–fluid interface*

The *no-slip boundary condition* states that a fluid element sticks to a solid surface. If $\mathbf{v}$ and $\mathbf{w}$ represent fluid and solid surface velocities, respectively, then the no-slip boundary condition is expressed as

$$\mathbf{v} \cdot \boldsymbol{\lambda} = \mathbf{w} \cdot \boldsymbol{\lambda} \quad \text{on a solid surface,} \qquad (1.3\text{-}1)$$

where $\boldsymbol{\lambda}$ is the unit tangent vector to the solid surface. Although this boundary condition has been successful in describing the characteristics of many types of flow, it fails in special circumstances, such as spreading of a liquid on a solid surface, corner flow, and extrusion of polymer melts from a capillary tube.

When phase transition does not take place between the phases, such as evaporation/condensation and melting/solidification, then there is no mass transfer across the interface. In this case, the normal components of the fluid and solid surface velocities must be equal to each other to satisfy the conservation of mass. Thus,

$$\mathbf{v} \cdot \mathbf{n} = \mathbf{w} \cdot \mathbf{n} \quad \text{on a solid surface,} \tag{1.3-2}$$

where $\mathbf{n}$ is the unit normal vector to the solid surface. Equation (1.3-2) is called the *kinematic boundary condition*.

Resolving $\mathbf{v}$ and $\mathbf{w}$ into their normal and tangential components gives

$$\mathbf{v} = (\mathbf{v} \cdot \mathbf{n}) \, \mathbf{n} + (\mathbf{v} \cdot \boldsymbol{\lambda}) \, \boldsymbol{\lambda}, \tag{1.3-3}$$

$$\mathbf{w} = (\mathbf{w} \cdot \mathbf{n}) \, \mathbf{n} + (\mathbf{w} \cdot \boldsymbol{\lambda}) \, \boldsymbol{\lambda}. \tag{1.3-4}$$

Subtraction of Eq. (1.3-4) from Eq. (1.3-3) leads to

$$\mathbf{v} - \mathbf{w} = \underbrace{\left[(\mathbf{v} - \mathbf{w}) \cdot \mathbf{n} \right]}_{0\,[\text{Eq. }(1.3-2)]} \mathbf{n} + \underbrace{\left[(\mathbf{v} - \mathbf{w}) \cdot \boldsymbol{\lambda} \right]}_{0\,[\text{Eq. }(1.3-1)]} \boldsymbol{\lambda}. \tag{1.3-5}$$

Therefore, Eq. (1.3-5) states that $\mathbf{v} = \mathbf{w}$ on a solid–fluid interface when there is no mass transfer between the phases.

1.3.2 *Conditions at a fluid–fluid interface*

Let us designate the two fluid phases in contact as α- and β-phases that are separated by a surface $A_{\alpha\beta} = A_{\beta\alpha}$. Let $\mathbf{n}_{\alpha\beta}$ be the unit vector to the surface $A_{\alpha\beta}$, pointing from the α-phase to the β-phase.

When there is no mass transfer between the α- and β-phases, then the normal components of the velocities of the α- and β-phases are continuous at the interface, i.e.,

$$\mathbf{v}_\alpha \cdot \mathbf{n}_{\alpha\beta} = \mathbf{v}_\beta \cdot \mathbf{n}_{\alpha\beta} \quad \text{on } A_{\alpha\beta}. \tag{1.3-6}$$

In addition, the no-slip boundary condition implies that

$$\mathbf{v}_\alpha \cdot \boldsymbol{\lambda} = \mathbf{v}_\beta \cdot \boldsymbol{\lambda} \quad \text{on } A_{\alpha\beta}, \tag{1.3-7}$$

where $\boldsymbol{\lambda}$ is a unit tangent vector to the interface. Therefore, $\mathbf{v}_\alpha = \mathbf{v}_\beta$ at the interface in the case of no mass transfer.

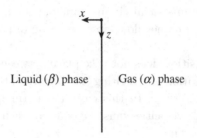

Figure 1.1   An interface between gas and liquid phases.

The jump momentum balance at the interface is expressed as

$$(P_\alpha - P_\beta)\, \mathbf{n}_{\alpha\beta} + (\boldsymbol{\tau}_\alpha - \boldsymbol{\tau}_\beta) \cdot \mathbf{n}_{\alpha\beta} = 2H\gamma\mathbf{n}_{\alpha\beta} - \nabla_s\gamma \quad \text{on } A_{\alpha\beta}, \quad (1.3\text{-}8)$$

where H is the mean curvature, γ is the interfacial tension, and ∇_s is the surface gradient. The interfacial tension is dependent on concentration and temperature. If it is uniform, then $\nabla_s\gamma = 0$ and Eq. (1.3-8) reduces to

$$(P_\alpha - P_\beta)\, \mathbf{n}_{\alpha\beta} + (\boldsymbol{\tau}_\alpha - \boldsymbol{\tau}_\beta) \cdot \mathbf{n}_{\alpha\beta} = 2H\gamma\mathbf{n}_{\alpha\beta} \quad \text{on } A_{\alpha\beta}. \quad (1.3\text{-}9)$$

Taking the dot product of Eq. (1.3-9) with $\boldsymbol{\lambda}$ gives[7]

$$\boldsymbol{\tau}_\alpha : \mathbf{n}_{\alpha\beta}\boldsymbol{\lambda} = \boldsymbol{\tau}_\beta : \mathbf{n}_{\alpha\beta}\boldsymbol{\lambda} \quad \text{on } A_{\alpha\beta}. \quad (1.3\text{-}10)$$

Equation (1.3-10) implies that the tangential stresses are continuous (or equal to each other) at the fluid–fluid interface.

Now let us consider a gas–liquid interface as shown in Figure 1.1. Since $\mathbf{n}_{\alpha\beta} = \mathbf{e}_x$ and $\boldsymbol{\lambda} = \mathbf{e}_z$, Equation (1.3-10) reduces to

$$\tau_{xz}^G = \tau_{xz}^L \quad \text{at } x = 0 \quad (1.3\text{-}11)$$

or

$$\mu_G \frac{dv_z^G}{dx} = \mu_L \frac{dv_z^L}{dx} \quad \text{at } x = 0 \quad (1.3\text{-}12)$$

Rearrangement of Eq. (1.3-12) gives

$$\frac{dv_z^L}{dx} = \frac{\mu_G}{\mu_L} \frac{dv_z^G}{dx} \quad \text{at } x = 0. \quad (1.3\text{-}13)$$

Since $\mu_G \ll \mu_L$, Eq. (1.3-13) simplifies to

$$\frac{dv_z^L}{dx} \simeq 0 \quad \text{at } x = 0. \quad (1.3\text{-}14)$$

---

[7]Keep in mind that $\boldsymbol{\lambda}$ is perpendicular to $\mathbf{n}_{\alpha\beta}$, i.e., $\boldsymbol{\lambda} \cdot \mathbf{n}_{\alpha\beta} = 0$.

Chapter 2

Fluid Statics

Two types of problems are encountered in fluid statics: (i) determination of hydrostatic pressure distribution within the fluid and (ii) calculation of the force of the fluid acting on submerged surfaces.

2.1 Hydrostatic Pressure Distribution in a Liquid

Consider a liquid contained in a tank open to the atmosphere as shown in Figure 2.1. Since the fluid is at rest, i.e., $\mathbf{v} = 0$, the equation of motion, Eq. (1.2-22), reduces to

$$0 = -\nabla P + \rho \mathbf{g}. \qquad (2.1\text{-}1)$$

The components of Eq. (2.1-1) in the x-, y-, and z-directions are given by

$$0 = -\frac{\partial P}{\partial x}, \qquad (2.1\text{-}2)$$

$$0 = -\frac{\partial P}{\partial y}, \qquad (2.1\text{-}3)$$

$$0 = -\frac{\partial P}{\partial z} - \rho g. \qquad (2.1\text{-}4)$$

Equations (2.1-2) and (2.1-3) indicate that pressure is uniform in the x- and y-directions, respectively. Since the liquid density is constant, Eq. (2.1-4) can be rearranged as

$$\int_{P_{\text{atm}}}^{P} dP = -\rho g \int_{H}^{z} dz. \qquad (2.1\text{-}5)$$

17

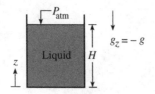

Figure 2.1   Static liquid in a tank.

Integration yields

$$P = P_{\text{atm}} + \rho g(H - z). \tag{2.1-6}$$

Therefore, the pressure, P, exerted by a column of liquid of height h is given by

$$P = \rho g h, \tag{2.1-7}$$

where $h = H - z$.

- The density of water is 1000 kg/m^3. Thus, the height of water that is equivalent to 1 atm (1.013×10^5 Pa) is

$$h = \frac{P}{\rho g} = \frac{1.013 \times 10^5}{(1000)(9.8)} = 10.3 \text{ m}.$$

As a rule of thumb, keep in mind that it takes 1 atm to move water up approximately 10 m.

- The density of blood is 1050 kg/m^3. Thus, the hydrostatic pressure difference from head to foot in a human 1.8 m tall is

$$\Delta P = \rho g h = (1050)(9.8)(1.8) \left(\frac{760}{1.013 \times 10^5} \frac{\text{mmHg}}{\text{Pa}} \right) \simeq 139 \text{ mmHg}.$$

Normal systolic blood pressure (maximum arterial pressure), however, is 120 mmHg due to the location of the heart.

2.2   Hydrostatic Pressure Distribution in Air

The highest point on the Earth's surface is the top of Mount Everest, which is approximately 8900 m above sea level. Let us estimate the air pressure at the top of Mount Everest with the following assumptions:

- air is stagnant,
- air is an ideal gas,

- the air temperature and pressure at sea level are 290 K and 1 atm, respectively,
- the rate of decrease in air temperature with altitude is 6.5 K/km.

Let z be the distance measured from sea level. The components of the equation of motion in the x-, y-, and z-directions are given by Eqs. (2.1-2), (2.1-3), and (2.1-4), respectively. Rearrangement of Eq. (2.1-4) gives

$$dP = -\rho g\, dz. \tag{2.2-1}$$

Since air is compressible, i.e., $\rho = \rho(T, P)$, Eq. (2.2-1) cannot be integrated directly. Therefore, it is first necessary to express density as a function of temperature and pressure. For an ideal gas

$$\rho = \frac{PM}{RT}, \tag{2.2-2}$$

where M is the molecular weight of air (29 kg/kmol) and R is the gas constant. The variation in temperature with elevation is expressed as

$$T = T_o - \beta z, \tag{2.2-3}$$

where $T_o = 290$ K and $\beta = 6.5 \times 10^{-3}$ K/m. Substitution of Eqs. (2.2-2) and (2.2-3) into Eq. (2.2-1) and rearrangement give

$$\int_{P_o}^{P} \frac{dP}{P} = -\frac{Mg}{R} \int_0^z \frac{dz}{T_o - \beta z}. \tag{2.2-4}$$

Integration of Eq. (2.2-4) results in

$$P = P_o \left(1 - \frac{\beta z}{T_o}\right)^{Mg/\beta R}. \tag{2.2-5}$$

Substitution of the numerical values yields

$$P = (1) \left[1 - \frac{(6.5)(8.9)}{290}\right]^{(29 \times 10^{-3})(9.8)/[(6.5 \times 10^{-3})(8.314)]} = 0.31 \text{ atm.}$$

At high altitudes, air molecules are separated from each other as a result of the decrease in atmospheric pressure, or in layman's terms, air becomes "thinner". Therefore, less oxygen enters the lungs, causing difficulty in breathing.

2.3   Forces on Submerged Bodies

Consider a unit normal vector drawn to a differential area dA, directed from the "negative side" of dA to the "positive side" of dA as shown in Figure 2.2. The vector $\mathbf{n} \cdot \boldsymbol{\pi} \, dA$ represents the force exerted by the fluid on the "negative side" of dA on the fluid on the "positive side" of dA, i.e.,

$$\mathbf{F} = \int_A \mathbf{n} \cdot \boldsymbol{\pi} \, dA. \qquad (2.3\text{-}1)$$

In the case of a static fluid, $\boldsymbol{\tau} = 0$ and $\boldsymbol{\pi} = P\mathbf{I}$. Thus, the force of the static fluid acting on an arbitrarily shaped solid with a surface area of A is given by

$$\mathbf{F} = \int_A \mathbf{n} P \, dA. \qquad (2.3\text{-}2)$$

According to the convention shown in Figure 2.2, the direction of the normal vector to the surface A should be from the fluid to the solid surface.[1] Note that Eq. (2.3-2) does not include any body forces, such as buoyancy force, acting on the solid.

Explanatory Example   Place your right hand parallel to the floor. The unit normal vector to the surface of your hand can be drawn in two ways.

(1) It may be directed from the top surface of your hand to the air (it is in the direction opposite to gravity). In this case, the "negative side" is the air adjacent to the lower face of your hand and the "positive side" is the air adjacent to the upper face of your hand. The vector $\mathbf{n} \cdot \pi dA$ is the force exerted by the air adjacent to the lower face on the air adjacent to the upper face.

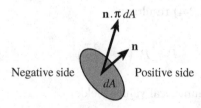

Figure 2.2   Differential surface area dA across which a force $\mathbf{n} \cdot \pi dA$ is transmitted (Bird *et al.*, 1987).

---

[1]If the direction of $\mathbf{n}$ is taken from the solid surface to the fluid, then $\mathbf{n}$ in Eq. (2.3-2) should be replaced by $-\mathbf{n}$.

(2) It may be directed from the lower surface of your hand to the air (it is in the direction of gravity). In this case, the "negative side" is the air adjacent to the upper face of your hand and the "positive side" is the air adjacent to the lower face of your hand. The vector $\mathbf{n} \cdot \boldsymbol{\pi} \, dA$ is the force exerted by the air adjacent to the upper face on the air adjacent to the lower face.

Note that the direction of the unit vector specifies the "negative" and "positive" sides. It is directed from the "negative" side to the "positive" side.

2.3.1 *Net force exerted on a vertical dam wall*

Let us calculate the net force exerted by water on a dam wall of height H and width W as shown in Figure 2.3. The hydrostatic pressure distribution is given by Eq. (2.1-6). Using Eq. (2.3-2), the net force acting on the wall is calculated from

$$\mathbf{F}_{\text{net}} = \underbrace{\int_0^W \int_0^H \mathbf{e}_x P \, dz \, dy}_{\text{Force exerted by water}} - \underbrace{\int_0^W \int_0^H \mathbf{e}_x P_{\text{atm}} \, dz \, dy}_{\text{Force exerted by air}}$$

$$= \mathbf{e}_x \int_0^W \int_0^H (P - P_{\text{atm}}) \, dz \, dy. \tag{2.3-3}$$

Substitution of Eq. (2.1-6) into Eq. (2.3-3) gives

$$\mathbf{F}_{\text{net}} = \mathbf{e}_x \int_0^W \int_0^H \rho g (H - z) \, dz \, dy = \frac{\rho g W H^2}{2} \mathbf{e}_x. \tag{2.3-4}$$

The x-component of the net force is

$$F_{\text{net},x} = \mathbf{F}_{\text{net}} \cdot \mathbf{e}_x = \frac{\rho g W H^2}{2}. \tag{2.3-5}$$

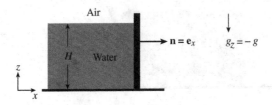

Figure 2.3   A vertical dam wall.

The average value of the pressure, $\langle P \rangle$, is given by

$$\langle P \rangle - P_{\text{atm}} = \frac{1}{H} \int_0^H \rho g (H - z)\, dz = \frac{\rho g H}{2}. \qquad (2.3\text{-}6)$$

Since the pressure distribution is linear, the average value occurs at the midpoint of the water height, $H/2$. The use of Eq. (2.3-6) in Eq. (2.3-5) leads to

$$F_{\text{net,x}} = \left(\frac{\rho g H}{2} \right)(WH) = (\langle P \rangle - P_{\text{atm}})\, A. \qquad (2.3\text{-}7)$$

Thus, multiplication of the average net pressure with area gives the net force in the x-direction.

2.3.2 *Net force exerted on an inclined dam wall*

Now let us calculate the net force exerted by water on an inclined dam wall making an angle α with the horizontal as shown in Figure 2.4. The width of the wall is W.

Using Eq. (2.3-2), the net force acting on the wall is calculated from

$$\mathbf{F}_{\text{net}} = \int_A \mathbf{n}\,(P - P_{\text{atm}})\, dA. \qquad (2.3\text{-}8)$$

The hydrostatic pressure distribution is given by Eq. (2.1-6). The components of the unit normal vector to the inclined dam wall in the x- and z-directions are expressed as

$$\mathbf{n} = (\mathbf{n} \cdot \mathbf{e}_x)\,\mathbf{e}_x + (\mathbf{n} \cdot \mathbf{e}_z)\,\mathbf{e}_z = \cos(90 - \alpha)\,\mathbf{e}_x - \cos\alpha\,\mathbf{e}_z$$
$$= \sin\alpha\,\mathbf{e}_x - \cos\alpha\,\mathbf{e}_z. \qquad (2.3\text{-}9)$$

The differential area of the wall is given by

$$dA = ds\, dy, \qquad (2.3\text{-}10)$$

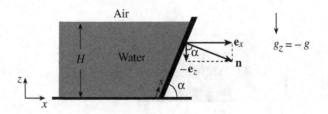

Figure 2.4   An inclined dam wall.

where s is the distance measured along the inclined wall. Note that

$$ds = \frac{dz}{\sin\alpha}. \tag{2.3-11}$$

Thus, Eq. (2.3-8) takes the form

$$\mathbf{F}_{\text{net}} = (\mathbf{e}_x - \cot\alpha\,\mathbf{e}_z) \int_0^W \int_0^H \rho g(H-z)\,dz\,dy$$

$$= \frac{\rho g W H^2}{2}\,(\mathbf{e}_x - \cot\alpha\,\mathbf{e}_z). \tag{2.3-12}$$

The x- and z-components of the net force are

$$F_{\text{net,x}} = \mathbf{F}_{\text{net}} \cdot \mathbf{e}_x = \frac{\rho g W H^2}{2}, \tag{2.3-13}$$

$$F_{\text{net,z}} = \mathbf{F}_{\text{net}} \cdot \mathbf{e}_z = -\frac{\rho g W H^2}{2}\cot\alpha. \tag{2.3-14}$$

The x-component of the net force, Eq. (2.3-13), is identical with Eq. (2.3-5). The term $WH^2\cot\alpha/2$ in Eq. (2.3-14) is the volume of the liquid above the inclined wall. Thus, the z-component of the net force is simply the weight of the fluid above the inclined wall.

The magnitude of the net resultant force is obtained from

$$F_{\text{net}} = \sqrt{F_{\text{net,x}}^2 + F_{\text{net,z}}^2}. \tag{2.3-15}$$

2.4 Projected Area Theorem

Force exerted by a static fluid on submerged objects can be calculated by Eq. (2.3-2). In Sections 2.3.1 and 2.3.2, the integral in Eq. (2.3-2) is evaluated by expressing the outwardly directed unit normal vector in terms of its components along the coordinate axes and the differential area in terms of the coordinates of the surface. This procedure, however, becomes very difficult for curved surfaces. An alternative procedure is to use the *projected area theorem* in the calculation of the force exerted by a static fluid. As shown in Figure 2.5, the projected area theorem expresses the projection of dA onto planes of constant x, y, and z by the following formulas:

The projection of dA on the y–z plane: $dA_x = \pm\,\mathbf{e}_x \cdot \mathbf{n}\,dA$ (2.4-1)

The projection of dA on the x–z plane: $dA_y = \pm\,\mathbf{e}_y \cdot \mathbf{n}\,dA$ (2.4-2)

The projection of dA on the x–y plane: $dA_z = \pm\,\mathbf{e}_z \cdot \mathbf{n}\,dA.$ (2.4-3)

The $\pm$ signs are chosen such that dA_x, dA_y, and dA_z are always positive.

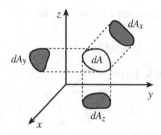

Figure 2.5   Projection of dA onto planes of constant x, y, and z.

2.4.1 *Net force exerted on an inclined dam wall: Revisited*

Let us resolve the example given in Section 2.3.2 using the projected area theorem. Taking the dot product of Eq. (2.3-8) with $\mathbf{e}_x$ and using Eq. (2.4-1) result in

$$F_{\text{net},x} = \int_A (P - P_{\text{atm}})\, dA_x = \int_0^W \int_0^H (P - P_{\text{atm}})\, dz\, dy. \qquad (2.4\text{-}4)$$

Substitution of Eq. (2.1-6) into Eq. (2.4-4) and integration lead to Eq. (2.3-13).

Taking the dot product of Eq. (2.3-8) with $\mathbf{e}_z$ and using Eq. (2.4-3) give the z-component of the net force as

$$F_{\text{net},z} = -\int_A (P - P_{\text{atm}})\, dA_z = -\int_A (P - P_{\text{atm}})\, dx\, dy. \qquad (2.4\text{-}5)$$

Substitution of Eq. (2.1-6) into Eq. (2.4-5) and noting

$$dx = \cot \alpha \, dz, \qquad (2.4\text{-}6)$$

result in

$$F_{\text{net},z} = -\int_0^W \int_0^H \cot \alpha \, \rho g (H - z)\, dz\, dy = -\frac{\rho g W H^2}{2} \cot \alpha, \qquad (2.4\text{-}7)$$

which is identical with Eq. (2.3-14).

2.4.2 *Net force exerted on a curved surface*

Let us calculate the horizontal and vertical components of the force per unit width exerted by water on the curved surface shown in Figure 2.6.

Using Eq. (2.3-2), the net force acting on the wall is calculated from

$$\mathbf{F}_{\text{net}} = \int_A \mathbf{n}\,(P - P_{\text{atm}})\, dA, \qquad (2.4\text{-}8)$$

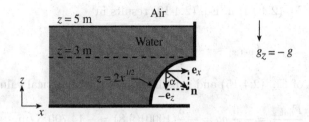

Figure 2.6   Force on a curved surface.

where

$$P - P_{\text{atm}} = \rho g(5 - z). \tag{2.4-9}$$

Substitution of Eq. (2.4-9) into Eq. (2.4-8) gives

$$\mathbf{F}_{\text{net}} = \rho g \int_A \mathbf{n}\,(5 - z)\, dA. \tag{2.4-10}$$

Taking the dot product of Eq. (2.4-10) with $\mathbf{e}_x$ and using Eq. (2.4-1) result in

$$F_{\text{net,x}} = \rho g \int_A (5 - z)\, dA_x = \rho g \int_0^W \int_0^3 (5 - z)\, dz\, dy. \tag{2.4-11}$$

Integration of Eq. (2.4-11) and substitution of the numerical values ($\rho = 1000\ \text{kg/m}^3$) yield

$$\frac{F_{\text{net,x}}}{W} = \frac{21}{2}\,\rho g = \frac{21}{2}(1000)(9.8) = 102{,}900\ \text{N/m}.$$

Taking the dot product of Eq. (2.4-10) with $\mathbf{e}_z$ and using Eq. (2.4-3) give the z-component of the net force as

$$F_{\text{net,z}} = -\rho g \int_A (5 - z)\, dA_z = -\rho g \int_A (5 - z)\, dx\, dy. \tag{2.4-12}$$

The equation of the curved surface is given by

$$z = 2\sqrt{x}. \tag{2.4-13}$$

Differentiation of Eq. (2.4-13) gives

$$dz = \frac{1}{\sqrt{x}}\, dx = \frac{2}{z}\, dx \quad \Rightarrow \quad dx = \frac{z}{2}\, dz. \tag{2.4-14}$$

The use of Eq. (2.4-14) in Eq. (2.4-12) results in

$$F_{net,z} = -\frac{\rho g}{2} \int_0^W \int_0^3 (5-z)z \, dz \, dy. \tag{2.4-15}$$

Integration of Eq. (2.4-15) and substitution of the numerical values yield

$$\frac{F_{net,z}}{W} = -\frac{3}{2}\rho g = -\frac{3}{2}(1000)(9.8) = -14{,}700 \text{ N/m}.$$

Chapter 3

Liquids in Rigid-Body Motion

A liquid in rigid-body motion moves without deformation as though it were a solid body. Since there is no deformation, there can be no shear stress, i.e., $\tau = 0$. Consequently, the only surface force on each element of liquid is that due to pressure.

3.1 Governing Equation

A liquid particle retains its identity in rigid-body motion because the fluid does not deform. In other words, deformation does not take place when each liquid particle in a container is subjected to the same acceleration vector, which is constant in time. Since $\tau = 0$, Eq. (1.2-17) simplifies to

$$\rho \, \mathbf{a} = - \nabla P + \rho \mathbf{g}, \tag{3.1-1}$$

where

$$\mathbf{a} = \frac{D\mathbf{v}}{Dt} = \frac{\partial \mathbf{v}}{\partial t} + \mathbf{v} \cdot \nabla \mathbf{v}. \tag{3.1-2}$$

3.2 Shape of a Liquid Surface in a Uniformly Accelerated Tank

A rectangular container partially filled with liquid at a depth of h_o is subjected to constant acceleration as shown in Figure 3.1. In the

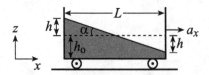

Figure 3.1 Constant acceleration of a tank partially filled with liquid on a straight path.

Cartesian coordinate system, the x-, y-, and z-components of Eq. (3.1-1) are expressed as

$$x\text{-component} \quad \rho a_x = -\frac{\partial P}{\partial x}, \tag{3.2-1}$$

$$y\text{-component} \quad 0 = \frac{\partial P}{\partial y}, \tag{3.2-2}$$

$$z\text{-component} \quad 0 = \frac{\partial P}{\partial z} + \rho g. \tag{3.2-3}$$

Equation (3.2-2) indicates that pressure does not depend on y. Therefore, the total differential of pressure is expressed as

$$dP = \left(\frac{\partial P}{\partial x}\right) dx + \left(\frac{\partial P}{\partial z}\right) dz. \tag{3.2-4}$$

Since the liquid surface is exposed to atmospheric pressure, $dP = 0$ along the surface. Substitution of Eqs. (3.2-1) and (3.2-3) into Eq. (3.2-4) yields

$$-\rho a_x \, dx - \rho g \, dz = 0 \quad \Rightarrow \quad \frac{dz}{dx} = -\frac{a_x}{g}. \tag{3.2-5}$$

Since dz/dx is constant, the free surface is a straight line and makes an angle α with the horizontal. The slope of the free surface is

$$\text{slope} = -\frac{a_x}{g} = -\tan\alpha = -\frac{h}{L/2} \quad \Rightarrow \quad h = \frac{L a_x}{2g}. \tag{3.2-6}$$

Thus, the heights of the free surface at $x = 0$ and $x = L$ are given as

$$z = h_o + \frac{L a_x}{2g} \quad \text{at } x = 0, \tag{3.2-7}$$

$$z = h_o - \frac{L a_x}{2g} \quad \text{at } x = L. \tag{3.2-8}$$

If the height of the tank is H, the maximum allowable acceleration to avoid spilling is given by

$$H = h_o + \frac{La_x}{2g} \quad \Rightarrow \quad a_x = \frac{2g(H - h_o)}{L}. \qquad (3.2\text{-}9)$$

Now consider a rectangular container partially filled with liquid at a depth of h_o undergoing constant acceleration along an inclined plane as shown in Figure 3.2. The components of the gravity vector are

$$g_x = g \sin 30, \qquad (3.2\text{-}10)$$

$$g_z = - g \cos 30. \qquad (3.2\text{-}11)$$

Thus, the x-, y-, and z-components of Eq. (3.1-1) are expressed as

$$x\text{-component} \quad \rho a_x = - \frac{\partial P}{\partial x} + \rho g \sin 30, \qquad (3.2\text{-}12)$$

$$y\text{-component} \quad 0 = \frac{\partial P}{\partial y}, \qquad (3.2\text{-}13)$$

$$z\text{-component} \quad 0 = \frac{\partial P}{\partial z} + \rho g \cos 30. \qquad (3.2\text{-}14)$$

Equation (3.2-13) indicates that pressure does not depend on y. Since $P = P(x, z)$, the total differential is expressed as

$$dP = \left(\frac{\partial P}{\partial x} \right) dx + \left(\frac{\partial P}{\partial z} \right) dz. \qquad (3.2\text{-}15)$$

Since the liquid surface is exposed to atmospheric pressure, $dP = 0$ along the surface. Substitution of Eqs. (3.2-12) and (3.2-14) into Eq. (3.2-15) yields

$$(-\rho a_x + \rho g \sin 30)\, dx - \rho g \cos 30\, dz = 0, \qquad (3.2\text{-}16)$$

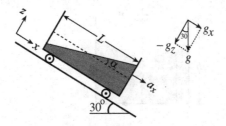

Figure 3.2   A container undergoing constant acceleration on an inclined plane.

or

$$\frac{dz}{dx} = \frac{g \sin 30 - a_x}{g \cos 30}. \tag{3.2-17}$$

Since dz/dx is constant, the free surface is a straight line with the slope

$$\text{slope} = \frac{g \sin 30 - a_x}{g \cos 30} = \tan \alpha. \tag{3.2-18}$$

If $a_x = 4 \text{ m/s}^2$, then

$$\tan \alpha = \frac{(9.8) \sin 30 - 4}{(9.8) \cos 30} = 0.106 \quad \Rightarrow \quad \alpha = 6.05°.$$

3.3   Shape of a Liquid Surface in a Uniformly Rotating Cylinder

A liquid of constant density and viscosity is in a cylindrical container of radius R. The container is caused to rotate about its own axis at an angular velocity Ω as shown in Figure 3.3. Find the shape of the free surface when steady-state has been established.

Since there is no flow either from the lateral surface or from the bottom of the tank, we postulate that $v_r = v_z = 0$ and $v_\theta = v_\theta(r)$. The use of Eq. (1.1-7) in Eq. (3.1-1) leads to the following equations:

$$r\text{-component} \quad \rho \frac{v_\theta^2}{r} = \frac{\partial P}{\partial r}, \tag{3.3-1}$$

$$\theta\text{-component} \quad 0 = \frac{\partial P}{\partial \theta}, \tag{3.3-2}$$

$$z\text{-component} \quad 0 = \frac{\partial P}{\partial z} + \rho g. \tag{3.3-3}$$

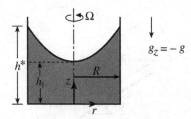

Figure 3.3   Liquid in a rotating cylindrical container.

To relate the linear velocity in the θ-direction, v_θ, to angular velocity, Ω, consider a particle moving along a circle of radius r as shown in Figure 3.4. When the displacement along the circumference is s, the corresponding angular displacement is θ (in radians). The relation between the linear and angular displacement is given by

$$s = r\theta. \tag{3.3-4}$$

The derivative of Eq. (3.3-4) with respect to time gives

$$\frac{ds}{dt} = r\frac{d\theta}{dt}, \tag{3.3-5}$$

or

$$v_\theta = r\Omega. \tag{3.3-6}$$

Thus, Eq. (3.3-1) becomes

$$\frac{\partial P}{\partial r} = \rho r \Omega^2. \tag{3.3-7}$$

Equation (3.3-2) indicates that pressure does not depend on θ. Thus, the total differential of pressure is expressed as

$$dP = \left(\frac{\partial P}{\partial r}\right) dr + \left(\frac{\partial P}{\partial z}\right) dz. \tag{3.3-8}$$

Since the free surface is exposed to the atmospheric pressure, then $dP = 0$ along the free surface. Substitution of Eqs. (3.3-7) and (3.3-3) into Eq. (3.3-8) yields

$$\rho \Omega^2 r\, dr - \rho g\, dz = 0. \tag{3.3-9}$$

Integration of Eq. (3.3-9) gives

$$\int_{h_1}^{z} dz = \frac{\Omega^2}{g} \int_0^r r\, dr, \tag{3.3-10}$$

Figure 3.4   Relation between angular and linear displacement.

or

$$z = h_1 + \frac{\Omega^2 r^2}{2g}, \qquad (3.3\text{-}11)$$

indicating that the equation of the free surface is a parabola with the vertex on the axis at $z = h_1$.

The height of the rim of the free surface, h^*, is calculated by evaluating Eq. (3.3-11) at $r = R$, i.e.,

$$h^* = h_1 + \frac{\Omega^2 R^2}{2g}. \qquad (3.3\text{-}12)$$

Since it is difficult to measure h_1, it is preferable to express z as a function of the initial height of the liquid when the container is stationary, h_o. Using the fact that the volume of the liquid in the container remains the same, one can write

$$\pi R^2 h_o = \int_0^R \int_0^z 2\pi r \, dz \, dr = 2\pi \int_0^R rz \, dr. \qquad (3.3\text{-}13)$$

Substitution of Eq. (3.3-11) into Eq. (3.3-13) and integration yield

$$h_1 = h_o - \frac{\Omega^2 R^2}{4g}. \qquad (3.3\text{-}14)$$

From Eqs. (3.3-12) and (3.3-14),

$$h^* = h_o + \frac{\Omega^2 R^2}{4g}. \qquad (3.3\text{-}15)$$

Numerical Example   Consider a vertical cylindrical container of 10 cm radius and 40 cm height partially filled with liquid. If the initial height of the liquid is 25 cm, the maximum rotational speed to avoid spilling from the edges of the container is calculated from Eq. (3.3-15) as

$$\Omega = \sqrt{\frac{4g(h^* - h_o)}{R^2}} = \sqrt{\frac{4(9.8)(0.4 - 0.25)}{(0.1)^2}} = 24.2 \text{ rad/s},$$

or

$$\Omega = \left(24.2\frac{\text{rad}}{\text{s}}\right)\left(60\frac{\text{s}}{\text{min}}\right)\left(\frac{1}{2\pi}\frac{\text{rev}}{\text{rad}}\right) = 231 \text{ rpm (rev/min)}.$$

The use of Eq. (3.3-14) in Eq. (3.3-11) results in

$$z = h_o - \frac{\Omega^2 R^2}{2g}\left[\frac{1}{2} - \left(\frac{r}{R}\right)^2\right]. \tag{3.3-16}$$

Note that Eq. (3.3-16) does not involve any liquid property. Thus, the shape of the free surface would be the same regardless of the liquid being used. Taking $R = 10$ cm and $h_o = 25$ cm, the variation in the shape of the free surface with rotational speed is shown in Figure 3.5.

To determine the pressure distribution, substitute Eqs. (3.3-3) and (3.3-7) into Eq. (3.3-8) to obtain

$$dP = \rho\Omega^2 r\,dr - \rho g\,dz. \tag{3.3-17}$$

Integration of Eq. (3.3-17) yields

$$P = \frac{\rho\Omega^2 r^2}{2} - \rho g z + C, \tag{3.3-18}$$

where C is an integration constant. The use of the boundary condition

$$\text{at } r = 0 \quad \text{and} \quad z = h_1 = h_o - \frac{\Omega^2 R^2}{4g} \qquad P = P_{\text{atm}} \tag{3.3-19}$$

gives

$$C = P_{\text{atm}} + \rho g\left(h_o - \frac{\Omega^2 R^2}{4g}\right). \tag{3.3-20}$$

Substitution of Eq. (3.3-20) into Eq. (3.3-18) gives

$$P = P_{\text{atm}} + \rho g h_o\left(1 - \frac{z}{h_o}\right) - \frac{\rho\Omega^2 R^2}{2}\left[\frac{1}{2} - \left(\frac{r}{R}\right)^2\right]. \tag{3.3-21}$$

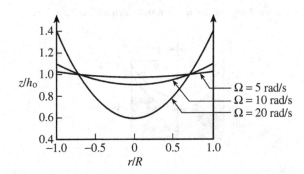

Figure 3.5   Variation in the shape of the free surface with rotational speed.

Since density appears in Eq. (3.3-21), pressure increases more rapidly with depth for a denser liquid. Figure 3.6 shows isobars for $R = 10$ cm, $h_o = 25$ cm, $\rho = 1000$ kg/m^3, and $\Omega = 15$ rad/s. Note that $P - P_{atm} = 0$ represents the free surface.

3.4   Shape of the Interface Between Two Immiscible Liquids in a Rotating Tank

In the previous section, it is found that a rotating cylindrical container containing a liquid results in a paraboloid of revolution for the free surface of the liquid. It is proposed that one can make plastic parabolic mirrors by pouring a molten epoxy resin onto the surface of a rotating mass of mercury as shown in Figure 3.7. Determine whether the mercury–plastic interface will indeed be parabolic. Assume no mixing between the phases and let superscripts I and II refer to mercury and plastic, respectively.

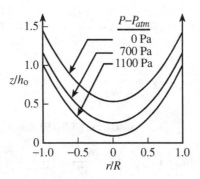

Figure 3.6   Isobars in a rotating cylindrical container.

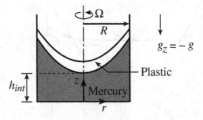

Figure 3.7   Two immiscible liquids in a rotating cylindrical container.

The governing equations for the mercury and plastic phases are shown in the following table:

	Mercury phase		Plastic phase	
r-component	$\dfrac{\partial P^{\mathrm{I}}}{\partial r} = \rho^{\mathrm{I}}\Omega^2 r$	(A)	$\dfrac{\partial P^{\mathrm{II}}}{\partial r} = \rho^{\mathrm{II}}\Omega^2 r$	(C)
z-component	$\dfrac{\partial P^{\mathrm{I}}}{\partial z} = -\rho^{\mathrm{I}}g$	(B)	$\dfrac{\partial P^{\mathrm{II}}}{\partial z} = -\rho^{\mathrm{II}}g$	(D)

Since $P^{\mathrm{I}} = P^{\mathrm{I}}(r, z)$, the total differential is expressed as

$$dP^{\mathrm{I}} = \left(\frac{\partial P^{\mathrm{I}}}{\partial r}\right) dr + \left(\frac{\partial P^{\mathrm{I}}}{\partial z}\right) dz. \tag{3.4-1}$$

Substitution of Eqs. (A) and (B) from the table into Eq. (3.4-1) gives

$$dP^{\mathrm{I}} = \rho^{\mathrm{I}}\Omega^2 r\, dr - \rho^{\mathrm{I}}g\, dz. \tag{3.4-2}$$

Integration of Eq. (3.4-2) yields

$$P^{\mathrm{I}}(r, z) = \frac{\rho^{\mathrm{I}}\Omega^2 r^2}{2} - \rho^{\mathrm{I}}gz + C_1, \tag{3.4-3}$$

where C_1 is an integration constant. Similarly, the pressure distribution in the plastic phase is given by

$$P^{\mathrm{II}}(r, z) = \frac{\rho^{\mathrm{II}}\Omega^2 r^2}{2} - \rho^{\mathrm{II}}gz + C_2. \tag{3.4-4}$$

Let h_{int} be the height of the mercury–plastic interface at $r = 0$. Application of the boundary condition

$$P^{\mathrm{I}}(0, h_{\mathrm{int}}) = P^{\mathrm{II}}(0, h_{\mathrm{int}}) = P_{\mathrm{int}}. \tag{3.4-5}$$

to Eqs. (3.4-3) and (3.4-4) yields

$$P_{\mathrm{int}} = -\rho^{\mathrm{I}}gh_{\mathrm{int}} + C_1 \quad \Rightarrow \quad C_1 = P_{\mathrm{int}} + \rho^{\mathrm{I}}gh_{\mathrm{int}}, \tag{3.4-6}$$

$$P_{\mathrm{int}} = -\rho^{\mathrm{II}}gh_{\mathrm{int}} + C_2 \quad \Rightarrow \quad C_2 = P_{\mathrm{int}} + \rho^{\mathrm{II}}gh_{\mathrm{int}}. \tag{3.4-7}$$

Therefore, Eqs. (3.4-6) and (3.4-7) take the form

$$P^{\mathrm{I}} = P_{\mathrm{int}} + \frac{\rho^{\mathrm{I}}\Omega^2 r^2}{2} - \rho^{\mathrm{I}}g(z - h_{\mathrm{int}}), \tag{3.4-8}$$

$$P^{\mathrm{II}} = P_{\mathrm{int}} + \frac{\rho^{\mathrm{II}}\Omega^2 r^2}{2} - \rho^{\mathrm{II}}g(z - h_{\mathrm{int}}). \tag{3.4-9}$$

At all points on the interface, $P^{\mathrm{I}} = P^{\mathrm{II}}$. Thus,

$$P_{\mathrm{int}} + \frac{\rho^{\mathrm{I}}\Omega^2 r^2}{2} - \rho^{\mathrm{I}}g(z - h_{\mathrm{int}}) = P_{\mathrm{int}} + \frac{\rho^{\mathrm{II}}\Omega^2 r^2}{2} - \rho^{\mathrm{II}}g(z - h_{\mathrm{int}}). \quad (3.4\text{-}10)$$

Simplification of Eq. (3.4-10) leads to

$$z = h_{\mathrm{int}} + \frac{\Omega^2 r^2}{2}. \quad (3.4\text{-}11)$$

Therefore, the mercury–plastic interface is indeed parabolic.

3.5   Physical Interpretation of the Modified Pressure

The modified pressure,[1] $\mathcal{P}$, is defined by Eq. (1.2-25) as

$$\mathcal{P} = P + \rho\widehat{\phi}, \quad (3.5\text{-}1)$$

where $\widehat{\phi}$ is the gravitational potential energy per unit mass. Let h be the distance measured in the direction opposite to gravity from any chosen reference plane. Thus, $\widehat{\phi} = gh$ and Eq. (3.5-1) becomes

$$\mathcal{P} = P + \rho gh. \quad (3.5\text{-}2)$$

Consider a stagnant liquid in a storage tank open to the atmosphere as shown in Figure 3.8. The hydrostatic pressure distribution within the fluid is given by

$$P = P_{\mathrm{atm}} + \rho gz. \quad (3.5\text{-}3)$$

The modified pressure, on the other hand, is defined as

$$\mathcal{P} = P - \rho gz. \quad (3.5\text{-}4)$$

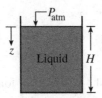

Figure 3.8   Static liquid in a tank.

---

[1]The term $\mathcal{P}$ is also called equivalent pressure, dynamic pressure, and piezometric pressure.

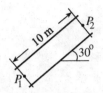

Figure 3.9   Flow in an inclined pipe.

Substitution of Eq. (3.5-3) into Eq. (3.5-4) gives

$$\mathcal{P} = P_{\text{atm}} = \text{constant.} \qquad (3.5\text{-}5)$$

The simplicity of defining the modified pressure comes from the fact that it is always constant under static conditions, whereas the hydrostatic pressure varies as a function of position. Suppose that you measure a pressure difference over a length L of a pipe. It is difficult to estimate whether this pressure difference comes from a flow situation or hydrostatic distribution. However, any variation in $\mathcal{P}$ implies a flow. Another distinct advantage of defining modified pressure is that the difference in $\mathcal{P}$ is independent of the orientation of the pipe.

Numerical Example   Water flows under steady conditions in an inclined pipe as shown in Figure 3.9. Determine the direction of flow if:

(a) $P_1 = 360$ kPa and $P_2 = 260$ kPa,
(b) $P_1 = 160$ kPa and $P_2 = 140$ kPa.

Solution

(a) Modified pressures at states 1 and 2 are

$$\mathcal{P}_1 = 360 \text{ kPa,}$$
$$\mathcal{P}_2 = 260 + \frac{(1000)\,(9.8)\,(10\sin 30)}{1000} = 309 \text{ kPa.}$$

Since $\mathcal{P}_1 > \mathcal{P}_2$, flow is from 1 to 2, i.e., upward.

(b) Modified pressures at states 1 and 2 are

$$\mathcal{P}_1 = 160 \text{ kPa,}$$
$$\mathcal{P}_2 = 140 + \frac{(1000)\,(9.8)\,(10\sin 30)}{1000} = 189 \text{ kPa.}$$

Since $\mathcal{P}_2 > \mathcal{P}_1$, flow is from 2 to 1, i.e., downward.

Chapter 4

Flow Due to Motion of Boundaries and/or Gravity

This chapter deals with fluid flow due to motion of boundaries[1] and/or gravity. The problems presented in this chapter will be analyzed with the following assumptions:

- steady-state,
- incompressible Newtonian fluid,
- one-dimensional, fully developed laminar flow,[2]
- constant physical properties.

4.1 Couette Flow Between Parallel Plates

Consider a Newtonian fluid between two parallel plates separated by distance H as shown in Figure 4.1. The width of the plate is W. The lower plate moves in the positive z-direction with a constant velocity of V, while the upper plate is held stationary. To determine the velocity distribution, functional forms of the non-zero velocity components should be postulated by making reasonable assumptions and examining the boundary conditions. Simplification of the velocity components is shown in Figure 4.2.[3]

---

[1]Flow generated by the movement of surface(s) enclosing the fluid is called *Couette flow*, named after the French physicist Maurice Marie Alfred Couette (1858–1943).

[2]*One-dimensional flow* implies only one non-zero velocity component. *Fully developed flow* implies no variation in velocity in the flow direction. Thus, the flow development regions near the entrance and exit are not taken into consideration, i.e., *end effects* are neglected.

[3]Fully developed flow implies $\partial v_z / \partial z = 0$. The same conclusion can also be reached from the equation of continuity, Eq. (A) in Table 1.1.

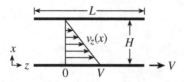

Figure 4.1   Couette flow between parallel plates.

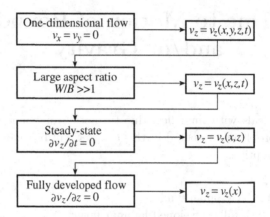

Figure 4.2   Simplification of the velocity components for Couette flow between parallel plates.

Taking $v_x = v_y = 0$ and $v_z = v_z(x)$, the components of the equation of motion, Eqs. (D), (E), and (F) in Table 1.5, simplify to

$$x\text{-component} \quad 0 = \frac{\partial P}{\partial x} + \rho g, \qquad (4.1\text{-}1)$$

$$y\text{-component} \quad 0 = \frac{\partial P}{\partial y}, \qquad (4.1\text{-}2)$$

$$z\text{-component} \quad 0 = \frac{d^2 v_z}{dx^2}. \qquad (4.1\text{-}3)$$

Equations (4.1-1) and (4.1-2) indicate that pressure depends only on x. Defining the modified pressure as

$$\mathcal{P} = P + \rho g x \qquad (4.1\text{-}4)$$

reduces Eq. (4.1-1) to

$$\frac{d\mathcal{P}}{dx} = 0 \quad \Rightarrow \quad \mathcal{P} = \text{constant}. \qquad (4.1\text{-}5)$$

Integration of Eq. (4.1-3) yields

$$v_z = C_1 x + C_2, \tag{4.1-6}$$

where C_1 and C_2 are constants of integration. The use of the boundary conditions

$$\text{at } x = 0 \quad v_z = V, \tag{4.1-7}$$
$$\text{at } x = H \quad v_z = 0, \tag{4.1-8}$$

gives the linear velocity distribution as

$$\frac{v_z}{V} = 1 - \frac{x}{H}. \tag{4.1-9}$$

The volumetric flow rate is obtained by integrating the velocity distribution over the cross-sectional area, i.e.,

$$Q = \int_0^W \int_0^H v_z \, dx \, dy. \tag{4.1-10}$$

Substitution of Eq. (4.1-9) into Eq. (4.1-10) and integration give

$$Q = \frac{HWV}{2}. \tag{4.1-11}$$

Dividing the volumetric flow rate by the flow area gives the average velocity as

$$\langle v_z \rangle = \frac{Q}{HW} = \frac{V}{2}. \tag{4.1-12}$$

From Eq. (F) in Table 1.2, the only non-zero shear stress component is

$$\tau_{xz} = -\mu \frac{dv_z}{dx} = \frac{\mu V}{H}, \tag{4.1-13}$$

which indicates that the shear stress is constant across the cross-section of the plate.

4.1.1 Drag force

A force exerted by a flowing fluid on the solid surfaces that it is in contact with is given by Eq. (2.3-1), i.e.,

$$\mathbf{F} = \int_A \mathbf{n} \cdot \boldsymbol{\pi} \, dA = \int_A \mathbf{n} P \, dA + \int_A \mathbf{n} \cdot \boldsymbol{\tau} \, dA, \tag{4.1-14}$$

where the direction of $\mathbf{n}$ is from the fluid to the solid surface. The *drag force*, F_D, is the force exerted by the fluid on the solid surface in the direction of mean flow. If $\boldsymbol{\lambda}$ is the unit vector in the direction of mean flow, then

$$F_D = \boldsymbol{\lambda} \cdot \mathbf{F} = \underbrace{\int_A (\boldsymbol{\lambda} \cdot \mathbf{n}) \, P \, dA_w}_{\text{Form Drag}} + \underbrace{\int_A \boldsymbol{\lambda} \cdot (\mathbf{n} \cdot \boldsymbol{\tau}) \, dA_w}_{\text{Friction (Skin) Drag}}. \qquad (4.1\text{-}15)$$

The integrands in Eq. (4.1-15) should be evaluated on the solid surface.

On the upper plate $\mathbf{n} = \mathbf{e}_x$. Since $\boldsymbol{\lambda} = \mathbf{e}_z$, then

$$\boldsymbol{\lambda} \cdot (\mathbf{n} \cdot \boldsymbol{\tau}) = \mathbf{e}_z \cdot (\mathbf{e}_x \cdot \tau_{ij} \mathbf{e}_i \mathbf{e}_j) = \mathbf{e}_z \cdot \tau_{xj} \mathbf{e}_j = \tau_{xz}. \qquad (4.1\text{-}16)$$

Since $\boldsymbol{\lambda}$ is orthogonal to $\mathbf{n}$, form drag is zero and Eq. (4.1-15) reduces to

$$F_D|_{\text{top}} = \int_0^L \int_0^W \tau_{xz}|_{x=H} \, dy \, dz. \qquad (4.1\text{-}17)$$

Substitution of Eq. (4.1-13) into Eq. (4.1-17) and integration yield

$$F_D|_{\text{top}} = \frac{\mu L W V}{H}. \qquad (4.1\text{-}18)$$

In other words, this is the force that must be applied to the upper plate in the negative z-direction to keep the plate stationary.

On the lower plate $\mathbf{n} = -\mathbf{e}_x$. Thus,

$$F_D|_{\text{bottom}} = -\frac{\mu L W V}{H}, \qquad (4.1\text{-}19)$$

which is the same magnitude as $F_D|_{\text{top}}$ but in the opposite direction. The rate of work done by the lower plate is calculated from Eq. (1.2-26) as

$$\dot{W} = F_D|_{\text{bottom}} \, V = -\frac{\mu L W V^2}{H}. \qquad (4.1\text{-}20)$$

This work dissipates as heat, i.e., irreversible degradation of mechanical energy into thermal energy, and causes an increase in fluid temperature.

In most cases, however, the temperature rise as a result of viscous dissipation is very small and cannot be detected by ordinary measuring devices. Therefore, for all practical purposes the flow is assumed to be isothermal.

4.1.2 Couette flow of two immiscible liquids

Now let us consider the flow of two immiscible liquids between parallel plates as shown in Figure 4.3. While the top plate is stationary, the bottom plate moves with velocity V.

Using Eq. (4.1-6), velocity distributions in liquids A and B are expressed as

$$v_z^A = C_1^A x + C_2^A, \tag{4.1-21}$$

$$v_z^B = C_1^B x + C_2^B. \tag{4.1-22}$$

The boundary conditions are

$$\text{at } x = H \qquad v_z^A = 0, \tag{4.1-23}$$

$$\text{at } x = 0 \qquad v_z^B = V, \tag{4.1-24}$$

$$\text{at } x = H/2 \qquad v_z^A = v_z^B, \tag{4.1-25}$$

$$\text{at } x = H/2 \qquad \mu_A \frac{dv_z^A}{dx} = \mu_B \frac{dv_z^B}{dx}. \tag{4.1-26}$$

Note that Eqs. (4.1-23) and (4.1-24) imply no-slip boundary condition. Equations (4.1-25) and (4.1-26) indicate that velocities and shear stresses are continuous at the liquid–liquid interface, respectively. Application of the boundary conditions leads to the following velocity distributions:

$$\frac{v_z^A}{V} = \frac{2\mu_B}{\mu_A + \mu_B} \left(1 - \frac{x}{H} \right), \tag{4.1-27}$$

$$\frac{v_z^B}{V} = 1 - \left(\frac{2\mu_A}{\mu_A + \mu_B} \right) \frac{x}{H}. \tag{4.1-28}$$

The volumetric flow rate of liquid A is

$$Q_A = \int_0^W \int_{H/2}^H v_z^A \, dx \, dy, \tag{4.1-29}$$

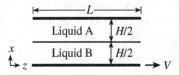

Figure 4.3   Couette flow of two immiscible liquids between parallel plates.

where W is the width of the plate. Similarly, the volumetric flow rate of liquid B is

$$Q_B = \int_0^W \int_0^{H/2} v_z^B \, dx \, dy. \tag{4.1-30}$$

Substitution of Eqs. (4.1-27) and (4.1-28) into Eqs. (4.1-29) and (4.1-30), respectively, and integration yield

$$Q_A = \frac{\mu_B HWV}{4(\mu_A + \mu_B)}, \tag{4.1-31}$$

$$Q_B = HWV \left[\frac{1}{2} - \frac{\mu_A}{4(\mu_A + \mu_B)} \right]. \tag{4.1-32}$$

The total volumetric flow rate is

$$Q = Q_A + Q_B = HWV \left[\frac{1}{2} + \frac{\mu_B - \mu_A}{4(\mu_A + \mu_B)} \right]. \tag{4.1-33}$$

The drag force exerted by liquid B on the surface of the bottom plate is

$$F_D|_{\text{bottom}} = -\int_0^L \int_0^W \tau_{xz}^B \big|_{x=0} \, dy \, dz = -\frac{2\mu_A \mu_B LWV}{(\mu_A + \mu_B)H}. \tag{4.1-34}$$

In the case of a single liquid phase, i.e., $\mu_A = \mu_B = \mu$, Eqs. (4.1-27) and (4.1-28) reduce to Eq. (4.1-9). On the other hand, Eqs. (4.1-33) and (4.1-34) reduce to Eqs. (4.1-11) and (4.1-19), respectively.

4.2   Couette Flow Between Rotating Concentric Cylinders

Consider a Newtonian fluid contained between two concentric cylinders of radii κR and R as shown in Figure 4.4. The inner and the outer cylinders rotate at angular velocities of Ω_i and Ω_o, respectively.

Simplification of the velocity components is shown in Figure 4.5.[4] When $L \gg \kappa R$, end effects at the cylinder bottom and at the free liquid interface can be ignored, i.e., $v_\theta \neq v_\theta(z)$. Taking $v_r = v_z = 0$ and $v_\theta = v_\theta(r)$, the components of the equation of motion, Eqs. (D), (E), and (F) in Table 1.6,

---

[4] Angular symmetry implies $\partial v_\theta / \partial \theta = 0$. The same conclusion can also be reached from the equation of continuity, Eq. (B) in Table 1.1.

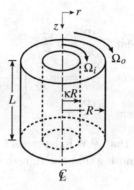

Figure 4.4   Flow between rotating concentric cylinders.

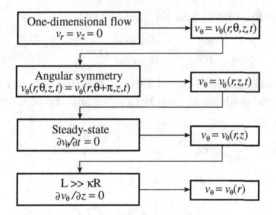

Figure 4.5   Simplification of the velocity components for flow between rotating concentric cylinders.

simplify to

r-component $\quad \rho \dfrac{v_\theta^2}{r} = \dfrac{\partial P}{\partial r},$ $\qquad$ (4.2-1)

θ-component $\quad 0 = -\dfrac{1}{r}\dfrac{\partial P}{\partial \theta} + \mu \dfrac{d}{dr}\left[\dfrac{1}{r}\dfrac{d}{dr}\left(rv_\theta\right)\right],$ $\qquad$ (4.2-2)

z-component $\quad 0 = -\dfrac{\partial P}{\partial z} + \rho g.$ $\qquad$ (4.2-3)

Defining the modified pressure as

$$\mathcal{P} = P - \rho g z.$$ $\qquad$ (4.2-4)

reduces Eq. (4.2-1)–(4.2-3) to

r-component $\quad \dfrac{\partial P}{\partial r} = \rho \dfrac{v_\theta^2}{r},$ (4.2-5)

θ-component $\quad \dfrac{\partial P}{\partial \theta} = \mu r \dfrac{d}{dr}\left[\dfrac{1}{r}\dfrac{d}{dr}(rv_\theta)\right],$ (4.2-6)

z-component $\quad \dfrac{\partial P}{\partial z} = 0.$ (4.2-7)

Equation (4.2-7) indicates that P is independent of z. Hence, $P = P(r,\theta)$. Integration of Eq. (4.2-5), noting that the right-hand side is dependent only on r, gives

$$P = \rho \int \frac{v_\theta^2}{r}\,dr + h(\theta).$$ (4.2-8)

Substitution of Eq. (4.2-8) into Eq. (4.2-6) yields

$$\frac{dh}{h\theta} = \mu r \frac{d}{dr}\left[\frac{1}{r}\frac{d}{dr}(rv_\theta)\right].$$ (4.2-9)

Since the right-hand side of Eq. (4.2-9) is a function only of r and the left-hand side depends only on θ, then both sides must be equal to a constant, say C_1. Thus,

$$\frac{dh}{h\theta} = C_1 \quad \Rightarrow \quad h = C_1\theta + C_2.$$ (4.2-10)

Substitution of Eq. (4.2-10) into Eq. (4.2-8) gives

$$P = \rho \int \frac{v_\theta^2}{r}\,dr + C_1\theta + C_2.$$ (4.2-11)

Since all quantities must be a periodic function of θ, i.e.,

$$P(r,\theta) = P(r,\theta + 2\pi) \quad \Rightarrow \quad C_1 = 0.$$ (4.2-12)

Therefore, pressure distribution becomes

$$P = \rho \int \frac{v_\theta^2}{r}\,dr + C_2,$$ (4.2-13)

where the integration constant C_2 can be evaluated by specifying the value of the modified pressure on the inner or outer cylinder.

Since $dh/d\theta$ is zero, Eq. (4.2-9) reduces to

$$\frac{d}{dr}\left[\frac{1}{r}\frac{d}{dr}(rv_\theta)\right] = 0.$$ (4.2-14)

The solution of Eq. (4.2-14) is

$$v_\theta = \frac{C_3}{2} r + \frac{C_4}{r} = C_5 r + \frac{C_4}{r}. \tag{4.2-15}$$

The boundary conditions are expressed as

$$\text{at } r = \kappa R \qquad v_\theta = \Omega_i \kappa R, \tag{4.2-16}$$

$$\text{at } r = R \qquad v_\theta = \Omega_o R. \tag{4.2-17}$$

Application of the boundary conditions yields the following equations:

$$\Omega_i \kappa R = C_5 \kappa R + \frac{C_4}{\kappa R}, \tag{4.2-18}$$

$$\Omega_o R = C_5 R + \frac{C_4}{R}. \tag{4.2-19}$$

In matrix notation, Eqs. (4.2-18) and (4.2-19) are expressed in the form of

$$\begin{pmatrix} \Omega_i \kappa R \\ \Omega_o R \end{pmatrix} = \begin{pmatrix} \kappa R & 1/\kappa R \\ R & 1/R \end{pmatrix} \begin{pmatrix} C_5 \\ C_4 \end{pmatrix}. \tag{4.2-20}$$

The constants are evaluated with the help of Cramer's rule as follows:

$$C_5 = \frac{\begin{vmatrix} \Omega_i \kappa R & 1/\kappa R \\ \Omega_o R & 1/R \end{vmatrix}}{\begin{vmatrix} \kappa R & 1/\kappa R \\ R & 1/R \end{vmatrix}} = \frac{\Omega_i \kappa - \Omega_o/\kappa}{\kappa - 1/\kappa} = \frac{\Omega_o - \Omega_i \kappa^2}{1 - \kappa^2}, \tag{4.2-21}$$

$$C_4 = \frac{\begin{vmatrix} \kappa R & \Omega_i \kappa R \\ R & \Omega_o R \end{vmatrix}}{\begin{vmatrix} \kappa R & 1/\kappa R \\ R & 1/R \end{vmatrix}} = \frac{\Omega_o \kappa R^2 - \Omega_i \kappa R^2}{\kappa - 1/\kappa} = -\frac{\kappa^2 R^2 (\Omega_o - \Omega_i)}{1 - \kappa^2}. \tag{4.2-22}$$

Substitution of Eqs. (4.2-21) and (4.2-22) into Eq. (4.2-15) and rearrangement result in

$$v_\theta = \frac{1}{R^2(1 - \kappa^2)} \left[(\Omega_o R^2 - \Omega_i \kappa^2 R^2) r - \frac{\kappa^2 R^4 (\Omega_o - \Omega_i)}{r} \right]. \tag{4.2-23}$$

Substitution of Eq. (4.2-23) into Eq. (4.2-13) and integration give the pressure distribution as

$$\mathcal{P} = \frac{\rho}{R^4(1 - \kappa^2)^2} \left[\frac{(\Omega_o R^2 - \Omega_i \kappa^2 R^2) r^2}{2} - \frac{\kappa^4 R^8}{2r^2} (\Omega_o - \Omega_i)^2 \right.$$

$$\left. - 2\kappa^2 R^4 (\Omega_o R^2 - \Omega_i \kappa^2 R^2)(\Omega_o - \Omega_i) \ln r \right] + C_2. \tag{4.2-24}$$

From Table 1.3, the only non-zero shear stress component is given by

$$\tau_{r\theta} = -\mu r \frac{d}{dr}\left(\frac{v_\theta}{r}\right). \tag{4.2-25}$$

The use of Eq. (4.2-23) in Eq. (4.2-25) gives the shear stress distribution as

$$\tau_{r\theta} = -\frac{2\mu\kappa^2 R^2 (\Omega_o - \Omega_i)}{1 - \kappa^2}\frac{1}{r^2}. \tag{4.2-26}$$

4.2.1   Calculation of torque

Torque, $\mathbf{T}$, is a vector whose direction is determined by the vector product of the position and force vectors, i.e.,

$$\mathbf{T} = \int_A \mathbf{r} \times (\mathbf{n} \cdot \boldsymbol{\pi})\, dA, \tag{4.2-27}$$

where the position vector, $\mathbf{r}$, is given by

$$\mathbf{r} = \begin{cases} x\,\mathbf{e}_x + y\,\mathbf{e}_y + z\,\mathbf{e}_z & \text{Cartesian coordinate} \\ r\,\mathbf{e}_r + z\,\mathbf{e}_z & \text{Cylindrical coordinate} \\ r\,\mathbf{e}_r & \text{Spherical coordinate.} \end{cases} \tag{4.2-28}$$

We are interested in the z-component of the torque. For this purpose, taking the scalar product of Eq. (4.2-27) with $\mathbf{e}_z$ gives

$$T_z = \mathbf{e}_z \cdot \mathbf{T} = \int_A \mathbf{e}_z \cdot [\mathbf{r} \times (\mathbf{n} \cdot \boldsymbol{\pi})]\, dA. \tag{4.2-29}$$

The integrand in Eq. (4.2-29) should be evaluated on the solid surface.

     To calculate the torque exerted by the cylinder on the fluid, the unit normal vector should be directed from the cylinder to the fluid.

• **Inner cylinder:** Since $\mathbf{n} = \mathbf{e}_r$, the integrand in Eq. (4.2-29) becomes

$$\begin{aligned} \mathbf{e}_z \cdot [\mathbf{r} \times (\mathbf{n} \cdot \boldsymbol{\pi})] &= \mathbf{e}_z \cdot [(r\,\mathbf{e}_r + z\,\mathbf{e}_z) \times (P\,\mathbf{e}_r + \mathbf{e}_r \cdot \tau_{ij}\mathbf{e}_i\mathbf{e}_j)] \\ &= \mathbf{e}_z \cdot [(r\,\mathbf{e}_r + z\,\mathbf{e}_z) \times (P\,\mathbf{e}_r + \tau_{rj}\mathbf{e}_j)] \\ &= \mathbf{e}_z \cdot (z\,P\,\mathbf{e}_\theta + \epsilon_{rjk}\,r\,\tau_{rj}\,\mathbf{e}_k + \epsilon_{zjk}\,z\,\tau_{rj}\,\mathbf{e}_k) \\ &= z\,P\,\underbrace{(\mathbf{e}_z \cdot \mathbf{e}_\theta)}_{0} + \underbrace{\epsilon_{rjz}\,r\,\tau_{rj}}_{j=\theta} + \underbrace{\epsilon_{zjz}\,z\,\tau_{rj}}_{0} \\ &= r\tau_{r\theta}. \end{aligned} \tag{4.2-30}$$

Note that the term A in Eq. (4.2-29) is the lateral surface of the rotating cylinder. Therefore, Eq. (4.2-29) simplifies to

$$T_z = \int_0^L \int_0^{2\pi} (r\tau_{r\theta})|_{r=\kappa R} \kappa R \, d\theta \, dz. \tag{4.2-31}$$

Substitution of Eq. (4.2-26) into Eq. (4.2-31) and integration give

$$T_z = -\frac{4\pi\mu L\kappa^2 R^2(\Omega_o - \Omega_i)}{1 - \kappa^2}. \tag{4.2-32}$$

• **Outer cylinder:** In this case $\mathbf{n} = -\mathbf{e}_r$. Thus, Eq. (4.2-29) becomes

$$T_z = -\int_0^L \int_0^{2\pi} (r\tau_{r\theta})|_{r=R} R \, d\theta \, dz. \tag{4.2-33}$$

Substitution of Eq. (4.2-26) into Eq. (4.2-33) and integration give

$$T_z = \frac{4\pi\mu L\kappa^2 R^2(\Omega_o - \Omega_i)}{1 - \kappa^2}. \tag{4.2-34}$$

4.2.2 *Outer cylinder is stationary*

When the outer cylinder (cup) is stationary while the inner cylinder (bob) is rotating with a constant angular velocity Ω, the resulting equipment is called a *cup-and-bob viscometer*. The viscosity of the fluid placed between concentric cylinders is determined by measuring the torque as a function of the rotational speed.

Setting $\Omega_o = 0$ and $\Omega_i = \Omega$, Eqs. (4.2-23), (4.2-24), and (4.2-32) simplify to the following form:

$$v_\theta = \frac{\Omega}{1 - \kappa^2}\left[\frac{(\kappa R)^2}{r} - \kappa^2 r\right], \tag{4.2-35}$$

$$\mathcal{P} = C_2 - \frac{\rho\Omega}{2(1 - \kappa^2)^2}\left[\frac{(\kappa R)^2 r^2}{R^4} + \frac{4(\kappa R)^4 \Omega \ln r}{R^2} + \frac{(\kappa R)^4 \Omega}{r^2}\right], \tag{4.2-36}$$

$$T_z = \frac{4\pi\mu L\Omega(\kappa R)^2}{1 - \kappa^2}. \tag{4.2-37}$$

Rearrangement of Eq. (4.2-37) yields

$$\frac{T_z}{4\pi LR^2}\left(\frac{1 - \kappa^2}{\kappa^2}\right) = \mu\Omega. \tag{4.2-38}$$

Therefore, if $T_z \left(1 - \kappa^2\right) / \left[4\pi L(\kappa R)^2\right]$ is plotted versus Ω at a specified temperature, the slope of the resulting straight line gives the viscosity of the fluid.

When the velocity of the inner rotating cylinder exceeds a critical value, flow will no longer be one-dimensional, i.e., the r- and z-components of the velocity are no longer zero. In other words, when the dimensionless Taylor number (ratio of centrifugal to viscous force) exceeds a critical value, flow becomes unstable. Different definitions used for the Taylor number in the literature are given by Schrimpf *et al.* (2021).

Numerical Example Consider a cup-and-bob viscometer with a height of 30 cm. The bob has an outer diameter of 20 cm and rotates at 200 rpm. The cup has an inner diameter of 20.4 cm. Calculate the viscosity of the liquid if the measured torque is 0.18 N · m.

Solution The ratio of the inner to outer radii (or diameters) is

$$\kappa = \frac{20}{20.4} = 0.9804$$

The angular velocity in rad/s is

$$\Omega = \left(200 \ \frac{\text{rev}}{\text{min}}\right)\left(\frac{1}{60} \ \frac{\text{min}}{\text{s}}\right)\left(2\pi \ \frac{\text{rad}}{\text{rev}}\right) = \frac{20\pi}{3} \ \text{rad/s}.$$

Substitution of the numerical values into Eq. (4.2-38) gives

$$\mu = \frac{0.18}{4\pi(0.30)(10.2 \times 10^{-2})^2}\left(\frac{1 - 0.9804^2}{0.9804^2}\right)\frac{1}{(20\pi/3)} = 8.849 \times 10^{-3} \ \text{Pa · s}.$$

The power required to rotate the inner cylinder at a specified angular velocity Ω (in rad/s) is calculated with the help of Eq. (1.2-26) as follows:

$$\text{Power} = (\text{Force})(\text{Velocity})$$
$$= \underbrace{(\text{Force})(\text{Distance})}_{\text{Torque}}(\text{Angular Velocity}), \qquad (4.2\text{-}39)$$

or

$$\dot{W} = T_z \, \Omega. \qquad (4.2\text{-}40)$$

4.2.2.1 *Investigation of limiting cases*

Once the solution to a given problem is obtained, it is a good practice to investigate the limiting cases, if possible, and compare the results with the known solutions. If the results match, this does not necessarily mean that the solution is correct; however, the chances that it is correct are fairly high.

Case (i) $R \to \infty$:
The fluid becomes unbounded when the outer radius goes to infinity, i.e., $R \to \infty$. As $R \to \infty$, the ratio of the inner to outer radii, κ, goes to zero. Under these circumstances, Eqs. (4.2-35)–(4.2-37) take the following form:

$$v_\theta = \frac{\Omega(\kappa R)^2}{r}, \tag{4.2-41}$$

$$P = C_2 - \frac{\rho(\kappa R)^4 \Omega^2}{2r^2}, \tag{4.2-42}$$

$$T_z = 4\pi\mu L\Omega(\kappa R)^2. \tag{4.2-43}$$

The integration constant C_2 in Eq. (4.2-42) is evaluated by the application of the boundary condition

$$\text{at } r = \infty \qquad P = P_\infty. \tag{4.2-44}$$

The result is

$$P = P_\infty - \frac{\rho(\kappa R)^4 \Omega^2}{2r^2}. \tag{4.2-45}$$

Note that P_∞, the pressure far from the rotating cylinder, is the maximum pressure; the minimum pressure occurs at the cylinder surface. This is why cavitation[5] often occurs near the tip of the impeller. The following condition should be satisfied to avoid cavitation:

$$P|_{r=\kappa R} \geq P^{\text{vap}}. \tag{4.2-46}$$

The use of Eq. (4.2-45) in Eq. (4.2-46) leads to

$$\Omega(\kappa R) \leq \sqrt{\frac{2\left(P_\infty - P^{\text{vap}}\right)}{\rho}}. \tag{4.2-47}$$

To quantify this result, let us consider water at atmospheric pressure and 293 K. Using

$$P_\infty = 1.013 \times 10^5 \text{ Pa} \qquad P^{\text{vap}} = 2.3 \times 10^3 \text{ Pa} \qquad \rho = 998 \text{ kg/m}^3,$$

---

[5]Cavitation is the formation of vapor bubbles when the pressure falls below the vapor pressure of the liquid. It causes erosion, vibration, and noise.

Eq. (4.2-47) results in

$$\Omega(\kappa R) \leq 14 \, \text{m/s},$$

which is often obtainable on a rotating shaft.

Case (ii) $\kappa \to 1$:
When the ratio of the inner to outer radii is close to unity, i.e., $\kappa \to 1$, a concentric annulus may be considered to be a thin-plane slit and its curvature can be neglected. As shown in Section 4.1, Couette flow between parallel plates results in a linear velocity distribution. Thus, Figure 4.6 approximates the velocity distribution as $\kappa \to 1$. Since the velocity gradient is constant, the shear stress is also constant and expressed in the following form:

$$\tau_{r\theta} = -\mu \frac{(0 - \Omega\kappa R)}{R(1 - \kappa)} = \frac{\mu\kappa\Omega}{1 - \kappa}. \tag{4.2-48}$$

Substitution of Eq. (4.2-48) into Eq. (4.2-31) and integration yield

$$T_z = 2\pi\mu LR^2\Omega \left(\frac{\kappa^3}{1 - \kappa}\right). \tag{4.2-49}$$

Let us resolve the numerical example given in Section 4.2.2 using Eq. (4.2-49). Substitution of the numerical values into Eq. (4.2-49) gives

$$\mu = \frac{0.18}{2\pi(0.30)(10.2 \times 10^{-2})^2} \frac{1}{(20\pi/3)} \left(\frac{1 - 0.9804}{0.9804^3}\right) = 9.115 \times 10^{-3} \, \text{Pa} \cdot \text{s},$$

which is 3% greater than the actual value.

4.3   Falling Liquid Film on a Vertical Plate

A liquid film of thickness δ flows down a vertical plate under the action of gravity as shown in Figure 4.7(a). Velocity components can be simplified

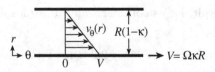

Figure 4.6   Approximation of the velocity distribution as $\kappa \to 1$.

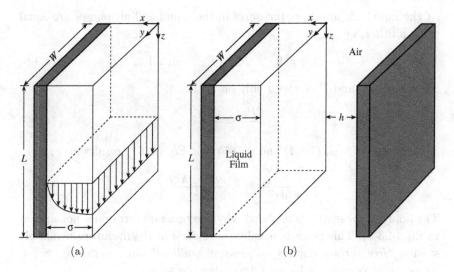

Figure 4.7   Falling liquid film on a vertical plate.

according to Figure 4.2.[6] Taking $v_x = v_y = 0$ and $v_z = v_z(x)$, the components of the equation of motion, Eqs. (D), (E), and (F) in Table 1.5, simplify to

$$x\text{-component} \quad 0 = \frac{\partial P_L}{\partial x}, \tag{4.3-1}$$

$$y\text{-component} \quad 0 = \frac{\partial P_L}{\partial y}, \tag{4.3-2}$$

$$z\text{-component} \quad 0 = -\frac{\partial P_L}{\partial z} + \mu_L \frac{d^2 v_z^L}{dx^2} + \rho_L g. \tag{4.3-3}$$

Equations (4.3-1) and (4.3-2) imply that $P_L \neq P_L(x)$ and $P_L \neq P_L(y)$, respectively. Since P_L depends only on z, Eq. (4.3-3) is expressed as

$$\frac{dP_L}{dz} = \mu_L \frac{d^2 v_z^L}{dx^2} + \rho_L g. \tag{4.3-4}$$

Now let us consider the z-component of the equation of motion for air. For a stagnant air phase, Eq. (4.3-4) is written as

$$\frac{dP_A}{dz} = \rho_A g. \tag{4.3-5}$$

---

[6]In this case, "large aspect ratio" implies $W/\delta \gg 1$.

At the liquid–air interface, pressures in the liquid and air phases are equal to each other, i.e.,

$$\text{at } x = 0 \qquad P_L = P_A \quad \text{for all } z. \tag{4.3-6}$$

Since both P_L and P_A depend only on z, then

$$\frac{dP_L}{dz} = \frac{dP_A}{dz}. \tag{4.3-7}$$

Substitution of Eqs. (4.3-4) and (4.3-5) into Eq. (4.3-7) results in

$$\frac{d^2 v_z^L}{dx^2} = -\frac{(\rho_L - \rho_A)g}{\mu^L}. \tag{4.3-8}$$

The analysis presented here shows why the pressure term does not appear in Eq. (4.3-8). This point is usually overlooked in the literature by simply stating *"free surface implies no pressure gradient"*. In general, $\rho_L \gg \rho_A$. Thus, $\rho_L - \rho_A \simeq \rho_L$ and Eq. (4.3-8) takes the form

$$\frac{d^2 v_z^L}{dx^2} = -\frac{\rho_L g}{\mu_L}. \tag{4.3-9}$$

Integration of Eq. (4.3-9) gives

$$\frac{d v_z^L}{dx} = -\frac{\rho_L g}{\mu_L} x + C_1. \tag{4.3-10}$$

Integration of Eq. (4.3-10) once more yields

$$v_z^L = -\frac{\rho_L g}{2\mu_L} x^2 + C_1 x + C_2. \tag{4.3-11}$$

The shear stress is continuous at the liquid–air interface, i.e.,

$$\text{at } x = 0 \qquad \tau_{xz}^L = \tau_{xz}^A \quad \Longrightarrow \quad \mu_L \frac{d v_z^L}{dx} = \mu_A \underbrace{\frac{d v_z^A}{dx}}_{0} \quad \Longrightarrow \quad \frac{d v_z^L}{dx} = 0$$
$$\tag{4.3-12}$$

The use of Eq. (4.3-12) in Eq. (4.3-10) gives $C_1 = 0$. The no-slip boundary condition on the wall surface is expressed as

$$\text{at } x = \delta \qquad v_z^L = 0. \tag{4.3-13}$$

Application of Eq. (4.3-13) to Eq. (4.3-11) and rearrangement result in

$$v_z^L = \frac{\rho_L g \delta^2}{2\mu_L} \left[1 - \left(\frac{x}{\delta} \right)^2 \right]. \tag{4.3-14}$$

A representative velocity distribution is also shown in Figure 4.7(a).[7] Once the velocity distribution is obtained, it is always a good habit to check whether it satisfies the boundary conditions.

The volumetric flow rate can be determined by integrating the velocity distribution across the flow area, i.e.,

$$Q = \int_0^W \int_0^\delta v_z \, dx \, dy. \tag{4.3-15}$$

Substitution of Eq. (4.3-14) into Eq. (4.3-15) and integration lead to

$$Q = \frac{\rho_L g \delta^3 W}{3\mu_L} \tag{4.3-16}$$

4.3.1 Falling liquid film confined between two vertical plates

Now let us repeat the analysis when another vertical plate is placed on the right-hand side of the falling liquid film as shown in Figure 4.7(b). In this case, air confined between the surface of the liquid and the plate placed on the right-hand side will be in motion.

The equation of motion for the liquid is given by Eq. (4.3-4), i.e.,

$$\frac{dP_L}{dz} = \mu_L \frac{d^2 v_z^L}{dx^2} + \rho_L g. \tag{4.3-17}$$

Similarly, the equation of motion for air is expressed as

$$\frac{dP_A}{dz} = \mu_A \frac{d^2 v_z^A}{dx^2} + \rho_A g. \tag{4.3-18}$$

Although $v_z^A \neq 0$, to simplify the analysis, the pressure gradient in air is assumed to be equal to the case when air is stagnant. In other words,

$$\frac{dP_A}{dz} = \rho_A g. \tag{4.3-19}$$

Since pressures in the liquid and air phases are equal to each other at the liquid–air interface, Eq. (4.3-7) is still valid. Thus, Eq. (4.3-17) takes the following form:

$$\frac{d^2 v_z^L}{dx^2} = -\frac{(\rho_L - \rho_A)g}{\mu_L} \simeq -\frac{\rho_L g}{\mu_L}. \tag{4.3-20}$$

---

[7]Students usually consider a representative velocity distribution to vary linearly from zero on the plate surface to some value on the surface of the liquid film, which is obviously not correct! A representative velocity distribution should be drawn to satisfy the boundary condition at the liquid–air interface, i.e., $dv_z/dx = 0$. In other words, the tangent drawn to the velocity distribution at $x = 0$ should be parallel to the x–y plane.

The use of Eq. (4.3-19) in Eq. (4.3-18) gives

$$\mu_A \frac{d^2 v_z^A}{dx^2} = 0. \tag{4.3-21}$$

Integration of Eq. (4.3-20) twice leads to

$$v_z^L = -\frac{\rho_L g}{2\mu_L} x^2 + C_1 x + C_2. \tag{4.3-22}$$

On the other hand, integration of Eq. (4.3-21) twice yields

$$v_z^A = \frac{C_3}{\mu_A} x + C_4. \tag{4.3-23}$$

The boundary conditions are

$$
\begin{aligned}
&\text{at } x = \delta & v_z^L &= 0, \\
&\text{at } x = -h & v_z^A &= 0, \\
&\text{at } x = 0 & v_z^L &= v_z^A, \\
&\text{at } x = 0 & \mu_L \frac{dv_z^L}{dx} &= \mu_A \frac{dv_z^A}{dx}.
\end{aligned} \tag{4.3-24}
$$

Application of the boundary conditions results in

$$v_z^L = \frac{\rho_L g \delta^2}{2\mu_L} \left[1 - \left(\frac{x}{\delta}\right)^2 - \frac{1 - (x/h)}{1 + (\mu_L h / \mu_A \delta)} \right], \tag{4.3-25}$$

$$v_z^A = \frac{\rho_L g \delta^2}{2\mu_L} \left[\frac{1 + (x/h)}{1 + (\mu_A \delta / \mu_L h)} \right]. \tag{4.3-26}$$

Note that Eq. (4.3-25) reduces to Eq. (4.3-14) when the third term in brackets is very small. In other words, the effect of the plate placed on the right-hand side of the falling film can be considered negligible if

$$\frac{\mu_L h}{\mu_A \delta} \gg 1 \quad \Rightarrow \quad \frac{h}{\delta} \gg \frac{\mu_A}{\mu_L}. \tag{4.3-27}$$

To quantify the criterion given by Eq. (4.3-27), let us assume that the liquid is water. The viscosities of air and water at 293 K are

$$\mu_A = 1.825 \times 10^{-5} \, \text{Pa} \cdot \text{s} \quad \text{and} \quad \mu_L = 1.002 \times 10^{-3} \, \text{Pa} \cdot \text{s}$$

Thus,

$$\frac{h}{\delta} \gg \frac{1.825 \times 10^{-5}}{1.002 \times 10^{-3}} = 0.018.$$

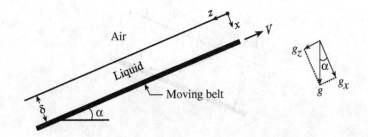

Figure 4.8   Flow of a liquid film on a moving belt.

4.4   Flow of a Liquid Film on a Moving Belt

A continuous wide belt, making an angle α with the horizontal, moves upward through a pool of liquid at a velocity V as shown in Figure 4.8. The thickness of the liquid film picked up by the belt is δ.

Velocity components can be simplified according to Figure 4.2.[8] Following the analysis presented in Section 4.3.1, the z-component of the equation of motion for the liquid film is expressed as

$$\frac{d^2 v_z}{dx^2} = -\frac{\rho g \sin \alpha}{\mu}. \tag{4.4-1}$$

Integration of Eq. (4.4-1) gives

$$\frac{dv_z}{dx} = -\frac{\rho g \sin \alpha}{\mu} x + C_1. \tag{4.4-2}$$

The use of the boundary condition

$$\text{at } x = 0 \qquad \frac{dv_z}{dx} = 0, \tag{4.4-3}$$

indicates that $C_1 = 0$. Integration of Eq. (4.4-3) once more leads to

$$v_z = -\frac{\rho g \sin \alpha}{2\mu} x^2 + C_2. \tag{4.4-4}$$

Application of the boundary condition on the belt surface, i.e.,

$$\text{at } x = \delta \qquad v_z = -V, \tag{4.4-5}$$

---

[8] In this case, "large aspect ratio" implies $W/\delta \gg 1$, where W is the width of the belt.

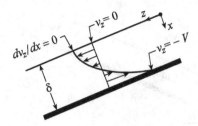

Figure 4.9   Representative velocity distribution in a film on a moving belt.

gives the velocity distribution as

$$v_z = \frac{\rho g \delta^2 \sin \alpha}{2\mu} \left[1 - \left(\frac{x}{\delta}\right)^2 \right] - V. \qquad (4.4\text{-}6)$$

A representative velocity distribution is shown in Figure 4.9. While gravity tends to pull the liquid downward, the movement of the belt drags the liquid upward.

The volumetric flow rate is

$$Q = \int_0^W \int_0^\delta v_z \, dx \, dy, \qquad (4.4\text{-}7)$$

Substitution of Eq. (4.4-6) into Eq. (4.4-7) and integration give

$$Q = \frac{W \rho g \delta^3 \sin \alpha}{3\mu} - W\delta V. \qquad (4.4\text{-}8)$$

When the volumetric flow rate is zero, Eq. (4.4-8) gives

$$V = \frac{\rho g \delta^2 \sin \alpha}{3\mu}. \qquad (4.4\text{-}9)$$

Substitution of Eq. (4.4-9) into Eq. (4.4-6) results in

$$v_z = \frac{\rho g \delta^2 \sin \alpha}{2\mu} \left[\frac{1}{3} - \left(\frac{x}{\delta}\right)^2 \right] \quad \text{when } Q = 0. \qquad (4.4\text{-}10)$$

Thus, when the net volumetric flow rate is zero

$$v_z > 0 \qquad \text{when } 0 \le x/\delta \le 1/3, \qquad (4.4\text{-}11\text{a})$$

$$v_z < 0 \qquad \text{when } 1/3 \le x/\delta \le 1. \qquad (4.4\text{-}11\text{b})$$

One can conclude from Eq. (4.4-8) that

$$\text{When } \frac{\rho g \delta^2 \sin \alpha}{3\mu} > V \quad \Rightarrow \quad Q > 0 \text{ (net downward flow)}, \qquad (4.4\text{-}12)$$

$$\text{When } \frac{\rho g \delta^2 \sin \alpha}{3\mu} < V \quad \Rightarrow \quad Q < 0 \text{ (net upward flow)}. \qquad (4.4\text{-}13)$$

4.5   Flow of Two Immiscible Liquids on an Inclined Plate

Consider flow of two immiscible liquids over an inclined plate under the action of gravity as shown in Figure 4.10. Velocity components can be simplified according to Figure 4.2. Taking $v_x = v_y = 0$ and $v_z = v_z(x)$, the components of the equation of motion, Eqs. (D), (E), and (F) in Table 1.5, simplify as follows:

Liquid A		Liquid B	
x-component			
$0 = \dfrac{\partial P_A}{\partial x} + \rho_A g \cos \alpha$	(A)	$0 = \dfrac{\partial P_B}{\partial x} + \rho_B g \cos \alpha$	(D)
y-component			
$0 = \dfrac{\partial P_A}{\partial y}$	(B)	$0 = \dfrac{\partial P_B}{\partial y}$	(E)
z-component			
$0 = -\dfrac{\partial P_A}{\partial z} + \mu_A \dfrac{d^2 v_z^A}{dx^2}$		$0 = -\dfrac{\partial P_B}{\partial z} + \mu_B \dfrac{d^2 v_z^B}{dx^2}$	
$\quad + \rho_A g \sin \alpha$	(C)	$\quad + \rho_B g \sin \alpha$	(F)

Integrations of Eqs. (A) and (D) yields

$$P_A = -\rho_A g \cos \alpha \, x + K_1, \tag{4.5-1}$$

$$P_B = -\rho_B g \cos \alpha \, x + K_2. \tag{4.5-2}$$

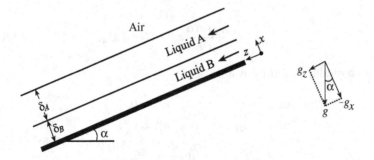

Figure 4.10   Flow of two immiscible liquids on an inclined plate.

Evaluations of K_1 and K_2 require the following boundary conditions:

$$\text{at } x = \delta_A + \delta_B \qquad P_A = P_{\text{atm}},$$
$$\text{at } x = \delta_B \qquad P_A = P_B. \tag{4.5-3}$$

Therefore, the pressure distributions become

$$P_A = P_{\text{atm}} + \rho_A g \cos \alpha \, (\delta_A + \delta_B - x), \tag{4.5-4}$$
$$P_B = P_{\text{atm}} + \rho_A g \delta_A \cos \alpha + \rho_B g \cos \alpha \, (\delta_B - x). \tag{4.5-5}$$

Since P_{atm} is constant, Eqs. (4.5-4) and (4.5-5) imply

$$\frac{\partial P_A}{\partial z} = \frac{\partial P_B}{\partial z} = 0, \tag{4.5-6}$$

so that the z-component of the equation of motion, Eqs. (C) and (F), simplify to

$$\frac{d^2 v_z^A}{dx^2} = -\frac{\rho_A g \sin \alpha}{\mu_A}, \tag{4.5-7}$$

$$\frac{d^2 v_z^B}{dx^2} = -\frac{\rho_B g \sin \alpha}{\mu_B}. \tag{4.5-8}$$

Integrations of Eqs. (4.5-7) and (4.5-8) twice lead to

$$v_z^A = -\frac{\rho_A g \sin \alpha}{2\mu_A} x^2 + C_1 x + C_2, \tag{4.5-9}$$

$$v_z^B = -\frac{\rho_B g \sin \alpha}{2\mu_B} x^2 + C_3 x + C_4. \tag{4.5-10}$$

The boundary conditions are

$$\text{at } x = 0 \qquad v_z^B = 0,$$
$$\text{at } x = \delta_B \qquad v_z^A = v_z^B,$$
$$\text{at } x = \delta_B \qquad \mu_A \frac{dv_z^A}{dx} = \mu_B \frac{dv_z^B}{dx},$$
$$\text{at } x = \delta_A + \delta_B \qquad \frac{dv_z^A}{dx} = 0. \tag{4.5-11}$$

Application of the boundary conditions leads to the following velocity distributions:

$$v_z^A = -\frac{\rho_A g \sin \alpha}{2\mu_A} x^2 + \frac{\rho_A g (\delta_A + \delta_B) \sin \alpha}{\mu_A} x + \frac{g \delta_B^2 \sin \alpha}{2} \left(\frac{\rho_B \mu_A - \rho_A \mu_B}{\mu_A \mu_B} \right)$$

$$+ \rho_A g \delta_A \delta_B \sin \alpha \left(\frac{\mu_A - \mu_B}{\mu_A \mu_B} \right), \tag{4.5-12}$$

$$v_z^B = -\frac{\rho_B g \sin \alpha}{2\mu_B} x^2 + \frac{(\rho_A \delta_A + \rho_B \delta_B) g \sin \alpha}{\mu_B} x. \tag{4.5-13}$$

The volumetric flow rate of liquid A is

$$Q_A = \int_0^W \int_{\delta_B}^{\delta_A + \delta_B} v_z^A \, dx \, dy, \tag{4.5-14}$$

where W is the width of the plate. Similarly, the volumetric flow rate of liquid B is

$$Q_B = \int_0^W \int_0^{\delta_B} v_z^B \, dx \, dy. \tag{4.5-15}$$

Substitution of Eqs. (4.5-12) and (4.5-13) into Eqs. (4.5-14) and (4.5-15), respectively, and integration yield

$$\frac{Q_A}{W} = -\frac{\rho_A g \sin \alpha}{6\mu_A} \left[(\delta_A + \delta_B)^3 - \delta_B^3 \right] + \frac{g \delta_A \delta_B^2 \sin \alpha}{2} \left(\frac{\rho_B \mu_A - \rho_A \mu_B}{\mu_A \mu_B} \right)$$

$$+ \frac{\rho_A g (\delta_A + \delta_B) \sin \alpha}{2\mu_A} \left[(\delta_A + \delta_B)^2 - \delta_B^2 \right]$$

$$+ \rho_A g \delta_A^2 \delta_B \sin \alpha \left(\frac{\mu_A - \mu_B}{\mu_A \mu_B} \right), \tag{4.5-16}$$

and

$$\frac{Q_B}{W} = -\frac{\rho_B g \delta_B^3 \sin \alpha}{6\mu_B} + \frac{(\rho_A \delta_A + \rho_B \delta_B) g \delta_B^2 \sin \alpha}{2\mu_B}. \tag{4.5-17}$$

4.5.1 *Investigation of a limiting case*

When a single liquid flows over an inclined plate, then

$$\rho_A = \rho_B = \rho \qquad \mu_A = \mu_B = \mu \qquad \delta_A + \delta_B = \delta. \tag{4.5-18}$$

Under these circumstances, summation of Eqs. (4.5-16) and (4.5-17) leads to

$$Q = Q_A + Q_B = \frac{\rho g \delta^3 W \sin \alpha}{3\mu}. \tag{4.5-19}$$

When $\alpha = 90°$, Eq. (4.5-19) is identical with Eq. (4.3-16).

4.6 Torque Required to Rotate a Cone

An incompressible Newtonian fluid is placed in the gap between coax-ial cones as shown in Figure 4.11(a). While the outer cone is stationary, the inner cone rotates with an angular velocity Ω. Determine the torque required if the difference between the cone angles, i.e., $\beta - \alpha = \epsilon$, is extremely small.

When the gap between the cones is small, the flow can be approximated by the Couette flow between parallel plates as shown in Figure 4.11(b). Calculation of torque requires the shear stress distribution to be known. From Eq. (4.1-13)

$$\tau_{xz} = \frac{\mu V}{H}. \tag{4.6-1}$$

Replacing τ_{xz} by $\tau_{\phi\theta}$, V by $\Omega r \sin \alpha$, and H by $r \sin \epsilon$, Eq. (4.6-1) becomes

$$\tau_{\phi\theta} = \frac{\mu \Omega \sin \alpha}{\sin \epsilon}. \tag{4.6-2}$$

Note that

$$\sin \epsilon \simeq \epsilon \qquad \text{when } \epsilon \text{ is small.} \tag{4.6-3}$$

Therefore, Eq. (4.6-2) simplifies to

$$\tau_{\phi\theta} = \frac{\mu \Omega \sin \alpha}{\epsilon}, \tag{4.6-4}$$

which is a constant.

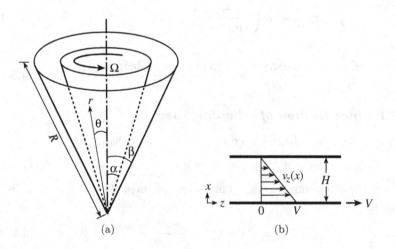

(a)   (b)

Figure 4.11 Flow between coaxial cones.

The torque required is given by

$$T_z = \int_0^{2\pi} \int_0^R \tau_{\phi\theta} \underbrace{(r\sin\alpha}_{\text{Lever arm}}) \underbrace{(r\sin\alpha \, dr d\phi)}_{\text{Eq. (A.6-14)}}. \tag{4.6-5}$$

Substitution of Eq. (4.6-4) into Eq. (4.6-5) and integration lead to

$$T_z = \frac{2\pi\mu\Omega(R\sin\alpha)^3}{3\epsilon}. \tag{4.6-6}$$

4.7 Flow in a Gear Pump

A schematic diagram of a gear pump is shown in Figure 4.12(a). Fluid is transported in the spaces between the teeth in the periphery of both gears.

Idealization of the flow between the teeth of the gear pump is shown in Figure 4.12(b), in which a fluid contained between a moving top surface and a rectangular cavity recirculates. Assume that $H \ll L$ so that the flow circulation near the left-hand and right-hand sides of the teeth is ignored and only the region where the flow is parallel to the top and bottom surfaces will be considered in the analysis.

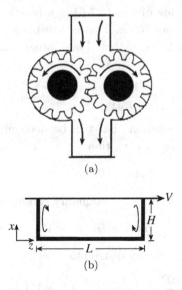

(a)

(b)

Figure 4.12   (a) A gear pump. (b) Idealization of a flow in a gear pump.

Postulating $v_x = v_y = 0$ and $v_z = v_z(x)$, the components of the equation of motion, Eqs. (D), (E), and (F) in Table 1.5, simplify to

$$x\text{-component} \quad 0 = \frac{\partial P}{\partial x} + \rho g, \tag{4.7-1}$$

$$y\text{-component} \quad 0 = \frac{\partial P}{\partial y}, \tag{4.7-2}$$

$$z\text{-component} \quad 0 = -\frac{\partial P}{\partial z} + \mu \frac{d^2 v_z}{dx^2}. \tag{4.7-3}$$

Equations (4.7-1)–(4.7-3) indicate that pressure depends on x and z. Defining the modified pressure as

$$\mathcal{P} = P + \rho g x \tag{4.7-4}$$

reduces Eq. (4.7-1) to

$$\frac{\partial \mathcal{P}}{\partial x} = 0 \quad \Rightarrow \quad \mathcal{P} \neq \mathcal{P}(x). \tag{4.7-5}$$

Since $\mathcal{P}$ is dependent only on z, Eq. (4.7-3) is rearranged as

$$\underbrace{\mu \frac{d^2 v_z}{dx^2}}_{f(x)} = \underbrace{\frac{d\mathcal{P}}{dz}}_{f(z)}. \tag{4.7-6}$$

While the right-hand side of Eq. (4.7-6) is a function of z only, the left-hand side is dependent only on x. This is possible if and only if both sides of Eq. (4.7-6) are equal to a constant. Thus, integration of Eq. (4.7-6) twice with respect to x yields

$$v_z = \frac{1}{2\mu} \frac{d\mathcal{P}}{dz} x^2 + C_1 x + C_2. \tag{4.7-7}$$

The term $d\mathcal{P}/dz$ is a constant that will be determined at a later stage. Application of the boundary condition

$$\text{at } x = 0 \quad v_z = 0, \tag{4.7-8}$$

gives $C_2 = 0$. The use of the other boundary condition, i.e.,

$$\text{at } x = H \quad v_z = V \tag{4.7-9}$$

results in the following velocity distribution:

$$v_z = -\frac{1}{2\mu} \frac{d\mathcal{P}}{dz} H^2 \left[\frac{x}{H} - \left(\frac{x}{H} \right)^2 \right] + V \left(\frac{x}{H} \right). \tag{4.7-10}$$

Assuming no leakage within the rectangular cavity, the net volumetric flow rate is zero, i.e.,

$$Q = \int_0^W \int_0^H v_z \, dx \, dy = W \int_0^H v_z \, dx = 0, \qquad (4.7\text{-}11)$$

where W is the width of the cavity. In terms of the dimensionless distance ξ, defined by

$$\xi = \frac{x}{H}, \qquad (4.7\text{-}12)$$

substitution of Eq. (4.7-10) into Eq. (4.7-11) gives

$$\frac{Q}{W} = 0 = -\frac{1}{2\mu} \frac{dP}{dz} H^3 \int_0^1 (\xi - \xi^2) \, d\xi + VH \int_0^1 \xi \, d\xi. \qquad (4.7\text{-}13)$$

Integration of Eq. (4.7-13) leads to

$$\frac{dP}{dz} = \frac{6\mu V}{H^2}. \qquad (4.7\text{-}14)$$

Therefore, the velocity distribution, Eq. (4.7-10), becomes

$$\frac{v_z}{V} = 3\left(\frac{x}{H}\right)^2 - 2\left(\frac{x}{H}\right). \qquad (4.7\text{-}15)$$

Figure 4.13 shows the velocity distribution based on Eq. (4.7-15).

From Eq. (F) in Table 1.2, the only non-zero shear stress component is

$$\tau_{xz} = -\mu \frac{dv_z}{dx}. \qquad (4.7\text{-}16)$$

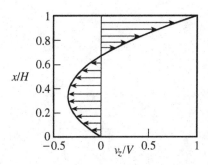

Figure 4.13 Velocity distribution in an idealized gear pump.

The use of Eq. (4.7-15) in Eq. (4.7-16) gives the shear stress distribution as

$$\tau_{xz} = \mu V \left(\frac{2}{H} - \frac{6x}{H^2} \right). \tag{4.7-17}$$

The drag force exerted by the fluid on the gear surface is given by Eq. (4.1-15), in which $\boldsymbol{\lambda} = \mathbf{e}_z$ and $\mathbf{n}$, the unit normal vector directed from the fluid to the gear surface, is $\mathbf{e}_x$. Thus, Eq. (4.1-15) simplifies to

$$F_D = \int_0^W \int_0^L \tau_{xz}|_{x=H} \, dz \, dy. \tag{4.7-18}$$

Substitution of Eq. (4.1-17) into Eq. (4.7-18) and integration lead to

$$F_D = -\frac{4\mu V L W}{H}. \tag{4.7-19}$$

Chapter 5

Flow Due to Pressure Gradient

This chapter deals with fluid flow due to imposed pressure gradient in the flow direction. The problems presented in this chapter will be analyzed with the following assumptions:

- steady-state,
- incompressible Newtonian fluid,
- one-dimensional, fully developed laminar flow,
- constant physical properties.

5.1 Flow Between Parallel Plates

Consider flow of a Newtonian fluid between two parallel plates as shown in Figure 5.1. A pressure gradient is imposed in the z-direction while both plates are held stationary.

Velocity components are simplified according to Figure 4.2. Taking $v_x = v_y = 0$ and $v_z = v_z(x)$, the components of the equation of motion, Eqs. (D), (E), and (F) in Table 1.5, simplify to

$$x\text{-component} \quad 0 = \frac{\partial P}{\partial x} + \rho g, \tag{5.1-1}$$

$$y\text{-component} \quad 0 = \frac{\partial P}{\partial y}, \tag{5.1-2}$$

$$z\text{-component} \quad 0 = -\frac{\partial P}{\partial z} + \mu \frac{d^2 v_z}{dx^2}. \tag{5.1-3}$$

Equation (5.1-2) indicates that $P \neq P(y)$. Thus, pressure is dependent on x and z. Rearrangement of Eq. (5.1-3) gives

$$\underbrace{\mu \frac{d^2 v_z}{dx^2}}_{f(x)} = \underbrace{\frac{\partial P}{\partial z}}_{f(x,z)} . \tag{5.1-4}$$

Since the dependence of P on x is not known, integration of Eq. (5.1-4) with respect to x is not possible at the moment. To circumvent this problem, let us define the modified pressure as

$$\mathcal{P} = P + \rho g x \tag{5.1-5}$$

so that

$$\frac{\partial \mathcal{P}}{\partial x} = \frac{\partial P}{\partial x} + \rho g, \tag{5.1-6}$$

and

$$\frac{\partial \mathcal{P}}{\partial z} = \frac{\partial P}{\partial z}. \tag{5.1-7}$$

Combination of Eqs. (5.1-1) and (5.1-6) yields

$$\frac{\partial \mathcal{P}}{\partial x} = 0, \tag{5.1-8}$$

which implies that $\mathcal{P} = \mathcal{P}(z)$ only. Therefore, the use of Eq. (5.1-7) in Eq. (5.1-4) gives

$$\underbrace{\mu \frac{d^2 v_z}{dx^2}}_{f(x)} = \underbrace{\frac{d\mathcal{P}}{dz}}_{f(z)} . \tag{5.1-9}$$

Note that while the right-hand side of Eq. (5.1-9) is a function of z only, the left-hand side is dependent only on x. This is possible if and only if both sides of Eq. (5.1-9) are equal to a constant, say λ. Hence,

$$\frac{d\mathcal{P}}{dz} = \lambda \quad \Rightarrow \quad \lambda = -\frac{\mathcal{P}_o - \mathcal{P}_L}{L}, \tag{5.1-10}$$

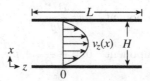

Figure 5.1   Flow between parallel plates.

where $\mathcal{P}_o$ and $\mathcal{P}_L$ are the values of $\mathcal{P}$ at $z = 0$ and $z = L$, respectively. Substitution of Eq. (5.1-10) into Eq. (5.1-9) gives the governing equation for velocity in the form

$$-\mu \frac{d^2 v_z}{dx^2} = \frac{\mathcal{P}_o - \mathcal{P}_L}{L}. \tag{5.1-11}$$

Integration of Eq. (5.1-11) twice results in

$$v_z = -\frac{\mathcal{P}_o - \mathcal{P}_L}{2\mu L} x^2 + C_1 x + C_2, \tag{5.1-12}$$

where C_1 and C_2 are integration constants. The use of the boundary conditions

$$\text{at } x = 0 \quad v_z = 0, \tag{5.1-13a}$$

$$\text{at } x = H \quad v_z = 0, \tag{5.1-13b}$$

gives the parabolic velocity distribution as

$$v_z = \frac{(\mathcal{P}_o - \mathcal{P}_L) H^2}{2\mu L} \left[\frac{x}{H} - \left(\frac{x}{H} \right)^2 \right]. \tag{5.1-14}$$

A representative plot of the velocity distribution is shown in Figure 5.1.

From Table 1.2, the only non-zero shear stress component is

$$\tau_{xz} = -\mu \frac{dv_z}{dx}. \tag{5.1-15}$$

Differentiation of Eq. (5.1-14) with respect to x and substitution of the resulting equation into Eq. (5.1-15) give the shear stress distribution as

$$\tau_{xz} = \frac{(\mathcal{P}_o - \mathcal{P}_L) H}{2L} \left[2 \left(\frac{x}{H} \right) - 1 \right]. \tag{5.1-16}$$

The volumetric flow rate can be determined by integrating the velocity distribution over the cross-sectional area, i.e.,

$$Q = \int_0^W \int_0^H v_z \, dx \, dy = W \int_0^H v_z \, dx, \tag{5.1-17}$$

where W is the width of the plate. Substitution of Eq. (5.1-14) into Eq. (5.1-17) and integration result in

$$Q = \frac{(\mathcal{P}_o - \mathcal{P}_L) W H^3}{12 \mu L}. \tag{5.1-18}$$

Dividing the volumetric flow rate by the flow area gives the average velocity as

$$\langle v_z \rangle = \frac{Q}{WH} = \frac{(\mathcal{P}_o - \mathcal{P}_L) H^2}{12 \mu L}. \tag{5.1-19}$$

5.1.1   *Calculation of drag force*

The drag force exerted by the fluid on the plate surfaces is given by Eq. (4.1-15), in which $\boldsymbol{\lambda} = \mathbf{e}_z$ and $\mathbf{n}$, the unit normal vector directed from the fluid to the plate surface, is given by

$$\mathbf{n} = \begin{cases} -\mathbf{e}_x & \text{at } x = 0, \\ \mathbf{e}_x & \text{at } x = H. \end{cases} \tag{5.1-20}$$

Since $\boldsymbol{\lambda}$ is orthogonal to $\mathbf{n}$, form drag is zero and Eq. (4.1-15) simplifies to

$$F_D = \int_0^L \int_0^W -\tau_{xz}\big|_{x=0} \, dy \, dz + \int_0^L \int_0^W \tau_{xz}\big|_{x=H} \, dy \, dz. \tag{5.1-21}$$

The use of Eq. (5.1-16) in Eq. (5.1-21) leads to

$$F_D = (\mathcal{P}_o - \mathcal{P}_L) WH. \tag{5.1-22}$$

The term on the right-hand side of Eq. (5.1-22) represents the summation of pressure and gravitational forces. Thus, Eq. (5.1-22) is simply Newton's second law of motion.

The power required to pump the fluid is calculated with the help of Eq. (1.2-26) as

$$\dot{W} = F_D \langle v_z \rangle = (\mathcal{P}_o - \mathcal{P}_L) WH \langle v_z \rangle = (\mathcal{P}_o - \mathcal{P}_L) \mathcal{Q}, \tag{5.1-23}$$

which is the product of pressure drop and volumetric flow rate.

5.1.2   *Friction factor*

The friction factor, f, is the dimensionless interaction of the system with the surroundings, defined by

$$F_D = A_{\mathrm{ch}} K_{\mathrm{ch}} f, \tag{5.1-24}$$

where A_{ch} is the characteristic area[1] and K_{ch} is the characteristic kinetic energy per unit volume. For flow between parallel plates,

$$A_{\mathrm{ch}} = 2WL \quad \text{and} \quad K_{\mathrm{ch}} = \frac{1}{2}\rho\langle v_z \rangle^2. \tag{5.1-25}$$

Elimination of the term $(\mathcal{P}_o - \mathcal{P}_L)$ between Eqs. (5.1-19) and (5.1-22) yields

$$F_D = \frac{12\mu LW \langle v_z \rangle}{H}. \tag{5.1-26}$$

---

[1] Characteristic area is the "wetted surface" for flow in conduits and the "projected area" for flow around submerged objects.

Substitution of Eqs. (5.1-25) and (5.1-26) into Eq. (5.1-24) and rearrangement result in

$$f = 12 \left(\frac{\mu}{H \langle v_z \rangle \rho} \right). \tag{5.1-27}$$

For flow in noncircular ducts, the Reynolds number, Re_h, based on the hydraulic equivalent diameter, D_h, is defined by

$$\text{Re}_h = \frac{D_h \langle v_z \rangle \rho}{\mu}, \tag{5.1-28}$$

where

$$D_h = 4 \left(\frac{\text{Flow area}}{\text{Wetted perimeter}} \right). \tag{5.1-29}$$

For flow between parallel plates,

$$D_h = 4 \left(\frac{WH}{2W} \right) = 2H. \tag{5.1-30}$$

Thus, Eq. (5.1-27) is expressed as

$$f = \frac{24}{\text{Re}_h}. \tag{5.1-31}$$

5.1.3 *Effect of using a different coordinate system*

Let us rework the problem using the coordinate system shown in Figure 5.2.

The velocity distribution is given by Eq. (5.1-12). Evaluation of the constants using the following boundary conditions:

$$\text{at } x = H/2 \qquad v_z = 0, \tag{5.1-32a}$$
$$\text{at } x = -H/2 \qquad v_z = 0, \tag{5.1-32b}$$

leads to

$$v_z = \frac{(\mathcal{P}_o - \mathcal{P}_L) H^2}{8 \mu L} \left[1 - 4 \left(\frac{x}{H} \right)^2 \right]. \tag{5.1-33}$$

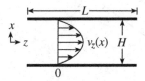

Figure 5.2   Flow between parallel plates.

Note that this result can be directly obtained from Eq. (5.1-14) by replacing x by $x + (H/2)$.

The volumetric flow rate is given by

$$Q = W \int_{-H/2}^{H/2} v_z \, dx. \tag{5.1-34}$$

Substitution of Eq. (5.1-33) into Eq. (5.1-34) and integration result in

$$Q = \frac{(\mathcal{P}_o - \mathcal{P}_L) W H^3}{12\mu L}, \tag{5.1-35}$$

which is identical with Eq. (5.1-18). Therefore, the volumetric flow rate is independent of the coordinate system.

5.2   Flow in a Circular Tube

A Newtonian fluid flows through an inclined tube of radius R as shown in Figure 5.3(a), in which the x-direction is parallel to the horizon.

Making use of the fact that vector quantities remain invariant under coordinate transformation, the components of the gravity vector are expressed as

$$g_r \mathbf{e}_r + g_\theta \mathbf{e}_\theta + g_z \mathbf{e}_z = g_x \mathbf{e}_x + g_y \mathbf{e}_y + g_z \mathbf{e}_z. \tag{5.2-1}$$

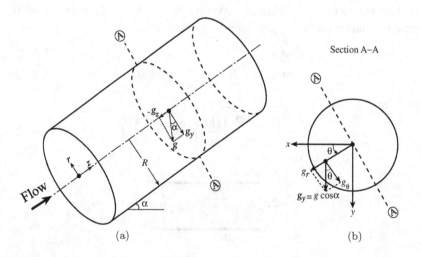

Figure 5.3   Flow in a circular tube.

Forming the scalar (or dot) product of Eq. (5.2-1) with $\mathbf{e}_r$ yields

$$g_r = g_x(\mathbf{e}_r \cdot \mathbf{e}_x) + g_y(\mathbf{e}_r \cdot \mathbf{e}_y). \tag{5.2-2}$$

Forming the scalar product of Eq. (5.2-1) with $\mathbf{e}_\theta$ gives

$$g_\theta = g_x(\mathbf{e}_\theta \cdot \mathbf{e}_x) + g_y(\mathbf{e}_\theta \cdot \mathbf{e}_y). \tag{5.2-3}$$

With respect to the Cartesian coordinate system shown in Figure 5.3(a), in which the x-direction is parallel to the horizon, the components of the gravity vector are

$$g_x = 0 \qquad g_y = g\cos\alpha \qquad g_z = -g\sin\alpha. \tag{5.2-4}$$

Substitution of Eq. (5.2-4) into Eqs. (5.2-2) and (5.2-3) results in

$$g_r = g\cos\alpha(\mathbf{e}_r \cdot \mathbf{e}_y), \tag{5.2-5}$$
$$g_\theta = g\cos\alpha(\mathbf{e}_\theta \cdot \mathbf{e}_y). \tag{5.2-6}$$

From Figure 5.3(b)

$$\mathbf{e}_r \cdot \mathbf{e}_y = \cos(90 - \theta) = \sin\theta \qquad \text{and} \qquad \mathbf{e}_\theta \cdot \mathbf{e}_y = \cos\theta \tag{5.2-7}$$

Substitution of Eq. (5.2-7) into Eqs. (5.2-5) and (5.2-6) gives the components of the gravity vector as

$$g_r = g\cos\alpha\sin\theta, \tag{5.2-8}$$
$$g_\theta = g\cos\alpha\cos\theta. \tag{5.2-9}$$

The simplification of the velocity components is shown in Figure 5.4.[2] Taking $v_r = v_\theta = 0$ and $v_z = v_z(r)$, the components of the equation of motion, Eqs. (D), (E), and (F) in Table 1.6, simplify to

$$r\text{-component} \quad 0 = -\frac{\partial P}{\partial r} + \rho g\cos\alpha\sin\theta, \tag{5.2-10}$$

$$\theta\text{-component} \quad 0 = -\frac{1}{r}\frac{\partial P}{\partial \theta} + \rho g\cos\alpha\cos\theta, \tag{5.2-11}$$

$$z\text{-component} \quad 0 = -\frac{\partial P}{\partial z} + \frac{\mu}{r}\frac{d}{dr}\left(r\frac{dv_z}{dr}\right) - \rho g\sin\alpha. \tag{5.2-12}$$

---

[2]Fully developed flow implies $\partial v_z/\partial z = 0$. The same conclusion can also be reached from the equation of continuity, Eq. (B) in Table 1.1.

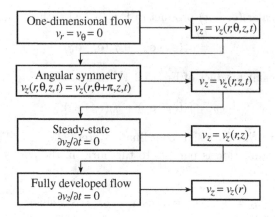

Figure 5.4   Simplification of the velocity components for flow through a circular tube.

Defining the modified pressure as

$$\mathcal{P} = P + \rho g z \sin\alpha, \qquad (5.2\text{-}13)$$

reduces Eqs. (5.2-10)–(5.2-12) to the following form:

$$\frac{\partial \mathcal{P}}{\partial r} = \rho g \cos\alpha \sin\theta, \qquad (5.2\text{-}14)$$

$$\frac{\partial \mathcal{P}}{\partial \theta} = \rho g r \cos\alpha \cos\theta, \qquad (5.2\text{-}15)$$

$$\frac{\partial \mathcal{P}}{\partial z} = \frac{\mu}{r}\frac{d}{dr}\left(r\frac{dv_z}{dr}\right). \qquad (5.2\text{-}16)$$

Integration of Eq. (5.2-14) with respect to r gives

$$\mathcal{P} = \rho g r \cos\alpha \sin\theta + f(\theta, z). \qquad (5.2\text{-}17)$$

Substitution of Eq. (5.2-17) into Eq. (5.2-15) yields

$$\frac{\partial f}{\partial \theta} = 0 \quad \Rightarrow \quad f \neq f(\theta) \quad \text{and} \quad f = f(z). \qquad (5.2\text{-}18)$$

Thus, Eq. (5.2-17) becomes

$$\mathcal{P} = \rho g r \cos\alpha \sin\theta + f(z). \qquad (5.2\text{-}19)$$

Substitution of Eq. (5.2-19) into Eq. (5.2-16) gives

$$\underbrace{\frac{df}{dz}}_{\text{function of } z} = \underbrace{\frac{\mu}{r}\frac{d}{dr}\left(r\frac{dv_z}{dr}\right)}_{\text{function of } r}. \qquad (5.2\text{-}20)$$

The term on the right-hand side of Eq. (5.2-20) is a function only of r, while the term on the left-hand side depends only on z. This is possible if both sides are equal to a constant, i.e.,

$$\frac{df}{dz} = A = \text{constant.} \qquad (5.2\text{-}21)$$

Integration of Eq. (5.2-21) and substitution of the resulting equation into Eq. (5.2-19) give

$$\mathcal{P} = \rho g r \cos\alpha \sin\theta + A z + B, \qquad (5.2\text{-}22)$$

where B is a constant of integration. Using the boundary conditions

$$\text{at } r = R; \quad \theta = 3\pi/2; \quad z = 0 \qquad \mathcal{P} = \mathcal{P}_o, \qquad (5.2\text{-}23)$$
$$\text{at } r = R; \quad \theta = 3\pi/2; \quad z = L \qquad \mathcal{P} = \mathcal{P}_L, \qquad (5.2\text{-}24)$$

the pressure distribution is obtained in the following form:

$$\mathcal{P} - \mathcal{P}_o = -\left(\frac{\mathcal{P}_o - \mathcal{P}_L}{L}\right) z + \rho g \cos\alpha (r \sin\theta + R). \qquad (5.2\text{-}25)$$

This type of rigorous analysis[3] is a prerequisite to obtain the velocity distribution using Eq. (5.2-16). Note that Eq. (5.2-16) cannot be integrated with respect to r unless the quantity $\partial\mathcal{P}/\partial z$ is either a constant or independent of r.

From Eq. (5.2-25),

$$\frac{\partial\mathcal{P}}{\partial z} = -\frac{\mathcal{P}_o - \mathcal{P}_L}{L} = \text{constant.} \qquad (5.2\text{-}26)$$

Thus, Eq. (5.2-16) takes the form

$$\frac{\mu}{r}\frac{d}{dr}\left(r\frac{dv_z}{dr}\right) = -\frac{\mathcal{P}_o - \mathcal{P}_L}{L}. \qquad (5.2\text{-}27)$$

Integration of Eq. (5.2-27) twice gives

$$v_z = -\frac{(\mathcal{P}_o - \mathcal{P}_L)}{4\mu L} r^2 + C_1 \ln r + C_2, \qquad (5.2\text{-}28)$$

---

[3]To avoid such analysis, gravity terms are usually considered negligible! The neglect of the gravity terms may lead to serious errors for flow of a dense fluid through a large pipeline. Moreover, if the gravity terms are considered negligible, then pressure varies only along the flow direction. From Eq. (5.2-25), this is the case when $\alpha = 90°$, i.e., a vertical pipe.

where C_1 and C_2 are constants of integration. The flow domain consists of the region between $r = 0$ and $r = R$. When $r = 0$, Eq. (5.2-28) indicates that $v_z = -\infty$. Therefore, C_1 must be zero to have a physically possible solution. This condition is usually expressed as "v_z is finite at $r = 0$". Alternatively, the symmetry condition implies that the maximum velocity occurs at the tube center and the use of the boundary condition

$$\text{at } r = 0 \qquad \frac{dv_z}{dr} = 0, \qquad (5.2\text{-}29)$$

also gives $C_1 = 0$. The no-slip boundary condition on the tube surface, i.e.,

$$\text{at } r = R \qquad v_z = 0, \qquad (5.2\text{-}30)$$

is used to evaluate the constant C_2 and the velocity distribution becomes

$$v_z = \frac{(\mathcal{P}_o - \mathcal{P}_L) R^2}{4\mu L}\left[1 - \left(\frac{r}{R}\right)^2\right]. \qquad (5.2\text{-}31)$$

A representative parabolic velocity distribution based on Eq. (5.2-31) is shown in Figure 5.5.

From Table 1.3, the only non-zero shear stress component is

$$\tau_{rz} = -\mu \frac{dv_z}{dr}. \qquad (5.2\text{-}32)$$

Differentiation of Eq. (5.2-31) with respect to r and substitution of the resulting equation into Eq. (5.2-32) give the shear stress distribution as

$$\tau_{rz} = \left(\frac{\mathcal{P}_o - \mathcal{P}_L}{2L}\right) r. \qquad (5.2\text{-}33)$$

The volumetric flow rate can be determined by integrating the velocity distribution over the cross-sectional area, i.e.,

$$Q = \int_0^{2\pi} \int_0^R v_z r \, dr \, d\theta = 2\pi \int_0^R v_z r \, dr. \qquad (5.2\text{-}34)$$

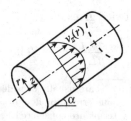

Figure 5.5   A representative velocity distribution for flow in a circular tube.

Substitution of Eq. (5.2-31) into Eq. (5.2-34) gives

$$Q = \frac{\pi \left(\mathcal{P}_o - \mathcal{P}_L\right) R^4}{8\mu L},$$

(5.2-35)

which is known as the *Hagen–Poiseuille law.*[4] Dividing the volumetric flow rate by the flow area gives the average velocity as

$$\langle v_z \rangle = \frac{Q}{\pi R^2} = \frac{\left(\mathcal{P}_o - \mathcal{P}_L\right) R^2}{8\mu L}.$$

(5.2-36)

5.2.1 Calculation of drag force

The drag force exerted by the fluid on the lateral surface of the tube is given by Eq. (4.1-15), in which $\boldsymbol{\lambda} = \mathbf{e}_z$ and $\mathbf{n}$, the unit normal vector directed from the fluid to the tube surface, is $\mathbf{e}_r$. Since $\boldsymbol{\lambda}$ is orthogonal to $\mathbf{n}$, form drag is zero and Eq. (4.1-15) simplifies to

$$F_D = \int_0^L \int_0^{2\pi} \tau_{rz}|_{r=R}\, R\, d\theta\, dz.$$

(5.2-37)

The use of Eq. (5.2-33) in Eq. (5.2-37) leads to

$$F_D = \left(\mathcal{P}_o - \mathcal{P}_L\right) \pi R^2.$$

(5.2-38)

The power required to pump the fluid is calculated with the help of Eq. (1.2-26) as

$$\dot{W} = F_D \langle v_z \rangle = \left(\mathcal{P}_o - \mathcal{P}_L\right) \pi R^2 \langle v_z \rangle = \left(\mathcal{P}_o - \mathcal{P}_L\right) Q,$$

(5.2-39)

which is the product of pressure drop and volumetric flow rate.

5.2.2 Friction factor

The characteristic area and the characteristic kinetic energy per unit volume are defined by

$$A_{ch} = \pi D L \quad \text{and} \quad K_{ch} = \frac{1}{2}\rho\langle v_z \rangle^2,$$

(5.2-40)

---

[4]Equation (5.2-35) was first obtained independently by a German hydraulic engineer Gotthilf Heinrich Ludwig Hagen in 1839 and a French physician Jean Léonard Marie Poiseuille in 1840. They formulated the equation after careful experiments using water in cylindrical tubes having different diameters and lengths.

where D is the tube diameter. Elimination of the term $(\mathcal{P}_o - \mathcal{P}_L)$ between Eqs. (5.2-35) and (5.2-38) yields

$$F_D = 8\pi\mu L\langle v_z\rangle. \tag{5.2-41}$$

Substitution of Eqs. (5.2-40) and (5.2-41) into Eq. (5.1-24) and rearrangement result in

$$f = 16\left(\frac{\mu}{D\langle v_z\rangle\rho}\right) = \frac{16}{\mathrm{Re}}, \tag{5.2-42}$$

where Re is the Reynolds number. The friction factor f appearing in Eq. (5.2-42) is called the *Fanning friction factor*. In the literature, another commonly used definition for the friction factor is the *Darcy friction factor*, f_D, which is four times greater than the Fanning friction factor, i.e.,

$$f_D = 4f. \tag{5.2-43}$$

Numerical Example   A Newtonian liquid is flowing in a horizontal pipe of diameter 0.02 m. The physical properties of the liquid are

$$\rho = 900 \text{ kg/m}^3 \qquad \text{and} \qquad \mu = 0.5 \text{ Pa}\cdot\text{s}.$$

(a) What should be the imposed pressure gradient in order to pump the liquid at a volumetric flow rate of 3×10^{-5} m^3/s?
(b) If the imposed pressure gradient is zero, i.e., $P_o = P_L$, what should be the angle of inclination of the pipe to achieve the same volumetric flow rate?

Solution   It is first necessary to check whether the flow is laminar. The Reynolds number is

$$\mathrm{Re} = \frac{D\langle v_z\rangle\rho}{\mu}.$$

Expressing the average velocity in terms of the volumetric flow rate

$$\langle v_z\rangle = \frac{Q}{\pi D^2/4},$$

reduces the expression for the Reynolds number to

$$\mathrm{Re} = \frac{4\rho Q}{\pi\mu D}.$$

Substitution of the numerical values gives

$$\text{Re} = \frac{4(900)(3 \times 10^{-5})}{\pi(0.5)(0.02)} = 3.44 < 2100 \quad \Rightarrow \quad \text{Laminar flow.}$$

(a) From Eq. (5.2-35),

$$\frac{\mathcal{P}_o - \mathcal{P}_L}{L} = \frac{8\mu Q}{\pi R^4} = \frac{8(0.5)(3 \times 10^{-5})}{\pi(0.01)^4} = 3820 \text{ Pa/m}.$$

From Eq. (5.2-13),

$$\frac{\mathcal{P}_o - \mathcal{P}_L}{L} = \frac{P_o - P_L}{L} - \rho g \sin \alpha.$$

Since $\alpha = 0°$ for a horizontal pipe, the pressure gradient is

$$\frac{P_o - P_L}{L} = \frac{\mathcal{P}_o - \mathcal{P}_L}{L} = 3820 \text{ Pa/m}.$$

(b) Since $P_o = P_L$,

$$\frac{\mathcal{P}_o - \mathcal{P}_L}{L} = -\rho g \sin \alpha = \frac{8\mu Q}{\pi R^4}.$$

Rearrangement yields

$$\sin \alpha = -\frac{8\mu Q}{\pi \rho g R^4} = -\frac{8(0.5)(3 \times 10^{-5})}{\pi(900)(9.8)(0.01)^4} = -0.433 \quad \Rightarrow \quad \alpha = -25.7°$$

5.3 Flow Through a Concentric Annulus

Consider a steady flow of a Newtonian fluid in an inclined concentric annulus as shown in Figure 5.6. A constant pressure gradient is imposed in the z-direction while the inner cylinder is stationary.

The development of the pressure and velocity distributions follow the same lines for flow in a circular tube with the following equations:

$$\mathcal{P} - \mathcal{P}_o = -\left(\frac{\mathcal{P}_o - \mathcal{P}_L}{L}\right) z + \rho g \cos \alpha (r \sin \theta + R), \tag{5.3-1}$$

and

$$v_z = -\frac{(\mathcal{P}_o - \mathcal{P}_L)}{4\mu L} r^2 + C_1 \ln r + C_2. \tag{5.3-2}$$

In this case, the flow domain consists of the region between $r = \kappa R$ and $r = R$. Since $r = 0$ is outside of the flow domain, Eq. (5.2-29) cannot be

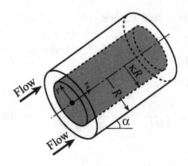

Figure 5.6   Flow through a concentric annulus.

used as a boundary condition. Application of the boundary condition on the outer tube surface, i.e.,

$$\text{at } r = R \qquad v_z = 0, \tag{5.3-3}$$

to Eq. (5.3-2) yields

$$C_2 = \frac{(\mathcal{P}_o - \mathcal{P}_L)\, R^2}{4\mu L} - C_1 \ln R. \tag{5.3-4}$$

Substitution of Eq. (5.3-4) into Eq. (5.3-2) and rearrangement give

$$v_z = \frac{(\mathcal{P}_o - \mathcal{P}_L)\, R^2}{4\mu L} \left[1 - \left(\frac{r}{R}\right)^2 \right] + C_1 \ln\left(\frac{r}{R}\right). \tag{5.3-5}$$

Application of the boundary condition on the inner tube surface, i.e.,

$$\text{at } r = \kappa R \qquad v_z = 0, \tag{5.3-6}$$

to Eq. (5.3-5) results in

$$C_1 = \frac{(\mathcal{P}_o - \mathcal{P}_L)\, R^2}{4\mu L} \left[\frac{1 - \kappa^2}{\ln(1/\kappa)} \right]. \tag{5.3-7}$$

Substitution of Eq. (5.3-7) into Eq. (5.3-5) leads to the following velocity distribution:

$$v_z = \frac{(\mathcal{P}_o - \mathcal{P}_L)\, R^2}{4\mu L} \left[1 - \left(\frac{r}{R}\right)^2 + \frac{(1 - \kappa^2)}{\ln(1/\kappa)} \ln\left(\frac{r}{R}\right) \right]. \tag{5.3-8}$$

From Table 1.3, the only non-zero shear stress component is

$$\tau_{rz} = -\mu \frac{dv_z}{dr}. \tag{5.3-9}$$

Differentiation of Eq. (5.3-8) with respect to r and substitution of the resulting equation into Eq. (5.3-9) give the shear stress distribution as

$$\tau_{rz} = \frac{(\mathcal{P}_o - \mathcal{P}_L)R}{2L}\left[\frac{r}{R} - \frac{(1-\kappa^2)}{2\ln(1/\kappa)}\frac{R}{r}\right]. \tag{5.3-10}$$

The volumetric flow rate can be determined by integrating the velocity distribution over the annular cross-sectional area, i.e.,

$$Q = \int_0^{2\pi}\int_{\kappa R}^R v_z r\, dr d\theta = 2\pi \int_{\kappa R}^R v_z r\, dr. \tag{5.3-11}$$

Before substituting Eq. (5.3-8) into Eq. (5.3-11), introduction of the dimensionless distance ξ, defined by

$$\xi = r/R, \tag{5.3-12}$$

simplifies the resulting expression as

$$Q = \frac{(\mathcal{P}_o - \mathcal{P}_L)\pi R^4}{2\mu L}\int_\kappa^1\left[1 - \xi^2 + \frac{(1-\kappa^2)}{\ln(1/\kappa)}\ln\xi\right]\xi\, d\xi. \tag{5.3-13}$$

Integration of Eq. (5.3-13) leads to[5]

$$Q = \frac{\pi\left(\mathcal{P}_o - \mathcal{P}_L\right)R^4}{8\mu L}\left[1 - \kappa^4 - \frac{(1-\kappa^2)^2}{\ln(1/\kappa)}\right]. \tag{5.3-14}$$

Dividing the volumetric flow rate by the flow area gives the average velocity as

$$\langle v_z\rangle = \frac{Q}{\pi R^2(1-\kappa^2)} = \frac{(\mathcal{P}_o - \mathcal{P}_L)R^2}{8\mu L}\left[1 + \kappa^2 - \frac{1-\kappa^2}{\ln(1/\kappa)}\right]. \tag{5.3-15}$$

5.3.1 Calculation of drag force

The drag force exerted by the fluid on the inner and outer tube surfaces is given by Eq. (4.1-15), in which $\lambda = \mathbf{e}_z$ and $\mathbf{n}$, the unit normal vector directed from the fluid to the tube surface, is given by

$$\mathbf{n} = \begin{cases} -\mathbf{e}_r & \text{at } r = \kappa R, \\ \mathbf{e}_r & \text{at } r = R. \end{cases} \tag{5.3-16}$$

---

[5]Note that

$$\int x\ln x\, dx = \left[x^2(\ln x - 1/2)\right]/2.$$

Since λ is orthogonal to $\mathbf{n}$, form drag is zero and Eq. (4.1-15) simplifies to

$$F_D = \int_0^L \int_0^{2\pi} -\tau_{rz}|_{r=\kappa R}\, \kappa R\, d\theta\, dz + \int_0^L \int_0^{2\pi} \tau_{rz}|_{r=R}\, R\, d\theta\, dz. \quad (5.3\text{-}17)$$

Substitution of Eq. (5.3-10) into Eq. (5.3-17) and integration lead to

$$F_D = (\mathcal{P}_o - \mathcal{P}_L)\,\pi R^2 (1 - \kappa^2). \quad (5.3\text{-}18)$$

The use of Eq. (1.2-26) gives the power required to pump the fluid as

$$\dot{W} = F_D \langle v_z \rangle = (\mathcal{P}_o - \mathcal{P}_L)\,\pi R^2 (1 - \kappa^2)\langle v_z \rangle = (\mathcal{P}_o - \mathcal{P}_L)\,\mathcal{Q}, \quad (5.3\text{-}19)$$

which is the product of pressure drop and volumetric flow rate.

5.3.2 *Friction factor*

The characteristic area and the characteristic kinetic energy per unit volume are defined by

$$A_{\text{ch}} = 2\pi R L(1 + \kappa) \qquad \text{and} \qquad K_{\text{ch}} = \frac{1}{2}\rho \langle v_z \rangle^2. \quad (5.3\text{-}20)$$

Elimination of the term $(\mathcal{P}_o - \mathcal{P}_L)$ between Eqs. (5.3-15) and (5.3-18) yields

$$F_D = 8\pi \mu L \langle v_z \rangle \left[\frac{1 - \kappa^2}{1 + \kappa^2 - \dfrac{1 - \kappa^2}{\ln(1/\kappa)}} \right]. \quad (5.3\text{-}21)$$

Substitution of Eqs. (5.3-20) and (5.3-21) into Eq. (5.1-24) and rearrangement result in

$$f = \frac{8\mu}{R\langle v_z \rangle \rho} \frac{(1 - \kappa)}{\left[1 + \kappa^2 - \dfrac{1 - \kappa^2}{\ln(1/\kappa)} \right]}. \quad (5.3\text{-}22)$$

The hydraulic equivalent diameter is

$$D_h = 4\left[\frac{\pi R^2 (1 - \kappa^2)}{2\pi R(1 + \kappa)} \right] = 2R(1 - \kappa). \quad (5.3\text{-}23)$$

The Reynolds number based on the hydraulic equivalent diameter is

$$\text{Re}_h = \frac{2R(1 - \kappa)\langle v_z \rangle \rho}{\mu}. \quad (5.3\text{-}24)$$

Thus, Eq. (5.3-22) is expressed as

$$f = \frac{16}{\text{Re}_h} \frac{(1 - \kappa)^2}{\left[1 + \kappa^2 - \dfrac{1 - \kappa^2}{\ln(1/\kappa)} \right]}. \quad (5.3\text{-}25)$$

5.3.3 *Investigation of limiting cases*

Case (i) $\kappa \to 0$:

When the ratio of the inner to outer radii is close to zero, i.e., $\kappa \to 0$, a concentric annulus is considered a circular pipe. In this case, Eq. (5.3-14) becomes

$$Q = \frac{\pi \left(\mathcal{P}_o - \mathcal{P}_L\right) R^4}{8\mu L} \lim_{\kappa \to 0} \left[1 - \kappa^4 - \frac{(1 - \kappa^2)^2}{\ln(1/\kappa)}\right]. \tag{5.3-26}$$

Noting that $\ln(1/\kappa) = -\ln \kappa$ and $\ln(0) = -\infty$, Eq. (5.3-26) reduces to

$$Q = \frac{\pi \left(\mathcal{P}_o - \mathcal{P}_L\right) R^4}{8\mu L}, \tag{5.3-27}$$

which is identical with Eq. (5.2-35).

On the other hand, Eq. (5.3-25) becomes

$$f = \lim_{\kappa \to 0} \left\{ \frac{16}{\mathrm{Re}_h} \frac{(1 - \kappa)^2}{\left[1 + \kappa^2 - \dfrac{1 - \kappa^2}{\ln(1/\kappa)}\right]} \right\} = \frac{16}{\mathrm{Re}}, \tag{5.3-28}$$

which is identical with Eq. (5.2-42).

Case (ii) $\kappa \to 1$:

When the ratio of the inner to outer radii is close to unity, i.e., $\kappa \to 1$, a concentric annulus is considered to be a thin-plane slit and its curvature can be neglected. Approximation of the flow through a concentric annulus by the flow between parallel plates requires the width of the plate, W, and the spacing between the plates, H, to be defined as

$$W = \frac{2\pi R + 2\pi \kappa R}{2} = \pi R \left(1 + \kappa\right), \tag{5.3-29}$$

$$H = R - \kappa R = R \left(1 - \kappa\right). \tag{5.3-30}$$

Therefore, the product WH^3 is equal to

$$WH^3 = \pi R^4 (1 - \kappa^2)(1 - \kappa)^2 \quad \Longrightarrow \quad \pi R^4 = \frac{WH^3}{(1 - \kappa^2)\left(1 - \kappa\right)^2}. \tag{5.3-31}$$

Thus, Eq. (5.3-14) becomes

$$Q = \frac{(\mathcal{P}_o - \mathcal{P}_L) WH^3}{8\,\mu L} \lim_{\kappa \to 1} \left[\frac{1 + \kappa^2}{(1 - \kappa)^2} + \frac{1 + \kappa}{(1 - \kappa)\ln \kappa}\right]. \tag{5.3-32}$$

Substitution of $\kappa = 1 - \varphi$ into Eq. (5.3-32) gives

$$Q = \frac{(\mathcal{P}_o - \mathcal{P}_L)\,WH^3}{8\,\mu L} \lim_{\varphi \to 0} \left[\frac{\varphi^2 - 2\varphi + 2}{\varphi^2} + \frac{2 - \varphi}{\varphi \ln(1 - \varphi)} \right]. \qquad (5.3\text{-}33)$$

The Taylor series expansion of the term $\ln(1 - \varphi)$ is

$$\ln(1 - \varphi) = -\varphi - \frac{1}{2}\varphi^2 - \frac{1}{3}\varphi^3 - \cdots \qquad (5.3\text{-}34)$$

Substitution of Eq. (5.3-34) into Eq. (5.3-33) and carrying out the division up to four terms yield

$$Q = \frac{(\mathcal{P}_o - \mathcal{P}_L)\,WH^3}{8\mu L} \lim_{\varphi \to 0} \left[1 - \frac{2}{\varphi} + \frac{2}{\varphi^2} \right.$$
$$\left. + \left(-\frac{2}{\varphi^2} + \frac{2}{\varphi} - \frac{1}{3} - \frac{\varphi}{2} \right) \right], \qquad (5.3\text{-}35)$$

or

$$Q = \frac{(\mathcal{P}_o - \mathcal{P}_L)\,WH^3}{8\mu L} \lim_{\varphi \to 0} \left(\frac{2}{3} - \frac{\varphi}{2} \right) = \frac{(\mathcal{P}_o - \mathcal{P}_L)\,WH^3}{12\mu L}, \qquad (5.3\text{-}36)$$

which is identical with Eq. (5.1-18).

On the other hand, substitution of $\kappa = 1 - \varphi$ into Eq. (5.3-25) gives

$$f = \frac{16}{\mathrm{Re}_h} \lim_{\varphi \to 0} \left\{ \frac{\varphi^2}{2 - 2\varphi + \varphi^2 + \dfrac{2\varphi - \varphi^2}{\ln(1 - \varphi)}} \right\}. \qquad (5.3\text{-}37)$$

Substitution of Eq. (5.3-34) into Eq. (5.3-37) and carrying out the division up to four terms yield

$$f = \frac{16}{\mathrm{Re}_h} \lim_{\varphi \to 0} \left\{ \frac{\varphi^2}{2 - 2\varphi + \varphi^2 - (2 - 2\varphi + \varphi^2/3 + \varphi^3/2)} \right\} = \frac{24}{\mathrm{Re}_h}, \qquad (5.3\text{-}38)$$

which is identical with Eq. (5.1-31).

5.3.3.1 *The inverse problem*

It was shown that the flow through a concentric annulus may be regarded as the flow between two parallel plates when the gap between the cylinders is small. Now let us consider the reverse problem, i.e., if we are given the velocity distribution, Eq. (5.1-14), and the volumetric flow rate, Eq. (5.1-18), for the flow between two parallel plates, can we obtain corresponding equations for the flow through a concentric annulus?

Substitution of Eq. (5.3-30) into Eq. (5.1-14) gives[6]

$$v_{z,\text{app}} = \frac{(\mathcal{P}_o - \mathcal{P}_L)\,R^2(1-\kappa)}{2\mu L}\left[\frac{r}{R} - \kappa - \frac{1}{1-\kappa}\left(\frac{r}{R} - \kappa\right)^2\right]. \qquad (5.3\text{-}39)$$

Note that the exact velocity distribution is given by Eq. (5.3-8). In terms of the dimensionless quantities, defined by

$$v^* = \frac{4\mu L}{(\mathcal{P}_o - \mathcal{P}_L)\,R^2}\,v_z \qquad \text{and} \qquad \xi = \frac{r}{R}, \qquad (5.3\text{-}40)$$

Eqs. (5.3-8) and (5.3-39) become

$$v^*_{\text{exact}} = 1 - \xi^2 + \frac{(1-\kappa^2)\ln\xi}{\ln(1/\kappa)}, \qquad (5.3\text{-}41)$$

$$v^*_{\text{app}} = 2(1-\kappa)\left[\xi - \kappa - \frac{(\xi-\kappa)^2}{1-\kappa}\right]. \qquad (5.3\text{-}42)$$

Comparison of the exact and approximate velocities as a function of κ and ξ is shown in Table 5.1. The results indicate that $v^*_{\text{app}} \simeq v^*_{\text{exact}}$ when $\kappa \geq 0.6$.

On the other hand, substitution of Eq. (5.3-31) into Eq. (5.1-18) yields

$$\mathcal{Q}_{\text{app}} = \frac{\pi\,(\mathcal{P}_o - \mathcal{P}_L)\,R^4}{12\mu L}(1-\kappa^2)(1-\kappa)^2. \qquad (5.3\text{-}43)$$

The exact volumetric flow rate is given by Eq. (5.3-14). In terms of the dimensionless volumetric flow rate, defined by

$$\mathcal{Q}^* = \frac{8\mu L}{\pi\,(\mathcal{P}_o - \mathcal{P}_L)\,R^4(1-\kappa^2)}, \qquad (5.3\text{-}44)$$

Eqs. (5.3-14) and (5.3-43) are expressed as

$$\mathcal{Q}^*_{\text{exact}} = 1 + \kappa^2 - \frac{1-\kappa^2}{\ln(1/\kappa)}, \qquad (5.3\text{-}45)$$

$$\mathcal{Q}^*_{\text{app}} = \frac{2}{3}(1-\kappa)^2. \qquad (5.3\text{-}46)$$

As shown in Table 5.2, $\mathcal{Q}^*_{\text{app}} \simeq \mathcal{Q}^*_{\text{exact}}$ when $\kappa \geq 0.6$.

---

[6]Note that $x = r - \kappa R$.

Table 5.1 Comparison of the exact and approximate velocity distributions.

	$\kappa = 0.6$		$\kappa = 0.8$	
ξ	v^*_{exact}	v^*_{app}	v^*_{exact}	v^*_{app}
0.60	0	0	–	–
0.65	0.0378	0.0350	–	–
0.70	0.0631	0.0600	–	–
0.75	0.0771	0.0750	–	–
0.80	0.0804	0.0800	0	0
0.85	0.0739	0.0750	0.0153	0.0150
0.90	0.0580	0.0600	0.0200	0.0200
0.95	0.0332	0.0350	0.0147	0.0150
1.00	0	0	0	0

Table 5.2 Comparison of the exact and approximate volumetric flow rates.

κ	0.6	0.70	0.80	0.90
$\mathcal{Q}^*_{\text{exact}}$	0.1071	0.0601	0.0267	6.668×10^{-3}
$\mathcal{Q}^*_{\text{app}}$	0.1067	0.0600	0.0267	6.667×10^{-3}

5.4   Flow of Two Immiscible Liquids in a Pipe

Consider flow of two immiscible liquids in a cylindrical pipe of radius R. The cross-sectional area of the pipe is shown in Figure 5.7, in which fluid A is at the core and fluid B encapsulates fluid A. The fluids have the same density but different viscosities ($\mu_A > \mu_B$).

Velocity distributions for fluids A and B are given by Eq. (5.2-28), i.e.,

$$v_z^A = -\frac{(\mathcal{P}_o - \mathcal{P}_L)}{4\mu_A L} r^2 + C_1 \ln r + C_2, \qquad (5.4\text{-}1)$$

$$v_z^B = -\frac{(\mathcal{P}_o - \mathcal{P}_L)}{4\mu_B L} r^2 + C_3 \ln r + C_4. \qquad (5.4\text{-}2)$$

Since v_z^A is finite at the center of the tube, $C_1 = 0$. The no-slip boundary condition at the tube wall, i.e.,

$$\text{at } r = R \qquad v_z^B = 0, \qquad (5.4\text{-}3)$$

reduces Eq. (5.4-2) to

$$v_z^B = \frac{(\mathcal{P}_o - \mathcal{P}_L) R^2}{4\mu_B L}\left[1 - \left(\frac{r}{R}\right)^2\right] + C_3 \ln\left(\frac{r}{R}\right). \qquad (5.4\text{-}4)$$

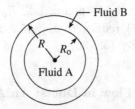

Figure 5.7　Flow of immiscible liquids in a pipe.

The boundary conditions at the liquid–liquid interface are

$$\text{at } r = R_o \qquad v_z^A = v_z^B, \qquad (5.4\text{-}5a)$$

$$\text{at } r = R_o \qquad \mu_A \frac{dv_z^A}{dx} = \mu_B \frac{dv_z^B}{dx}. \qquad (5.4\text{-}5b)$$

Therefore, the velocity distributions are expressed as

$$v_z^A = \frac{(P_o - P_L) R^2}{4\mu_B L} \left\{ 1 - \left(\frac{R_o}{R}\right)^2 + \frac{\mu_B}{\mu_A} \left[\left(\frac{R_o}{R}\right)^2 - \left(\frac{r}{R}\right)^2 \right] \right\}, \qquad (5.4\text{-}6)$$

$$v_z^B = \frac{(P_o - P_L) R^2}{4\mu_B L} \left[1 - \left(\frac{r}{R}\right)^2 \right]. \qquad (5.4\text{-}7)$$

The volumetric flow rates are

$$Q_A = \int_0^{2\pi} \int_0^{R_o} v_z^A r \, dr \, d\theta = 2\pi \int_0^{R_o} v_z^A r \, dr$$

$$= \frac{\pi (P_o - P_L) R^2 R_o^2}{4\mu_B L} \left[1 - \left(\frac{R_o}{R}\right)^2 + \frac{1}{2}\frac{\mu_B}{\mu_A} \left(\frac{R_o}{R}\right)^2 \right], \qquad (5.4\text{-}8)$$

$$Q_B = \int_0^{2\pi} \int_{R_o}^{R} v_z^B r \, dr \, d\theta = 2\pi \int_{R_o}^{R} v_z^B r \, dr$$

$$= \frac{\pi (P_o - P_L) R^2 R_o^2}{4\mu_B L} \left[\frac{1}{2} \left(\frac{R}{R_o}\right)^2 + \frac{1}{2} \left(\frac{R_o}{R}\right)^2 - 1 \right]. \qquad (5.4\text{-}9)$$

The total volumetric flow rate is given by

$$Q = Q_A + Q_B = \frac{\pi (P_o - P_L) R^2 R_o^2}{8\mu_B L} \left[\left(\frac{R_o}{R}\right)^2 \left(\frac{\mu_B}{\mu_A} - 1\right) + \left(\frac{R}{R_o}\right)^2 \right].$$

$$(5.4\text{-}10)$$

Taking $\mu_A/\mu_B = 100$ and $R_o/R = 0.8$, $Q_A/Q_B = 3.587$.

In the case of a single liquid phase, i.e., $R_o = R$ and $\mu_A = \mu_B = \mu$, Eqs. (5.4-6) and (5.4-7) reduce to Eq. (5.2-31). On the other hand, Eq. (5.4-10) simplifies to Eq. (5.2-35).

5.5   One-Dimensional Flow in Ducts: An Alternative Approach

Consider one-dimensional, laminar flow of an incompressible Newtonian fluid under steady conditions through a duct having an arbitrary but constant cross-sectional area as shown in Figure 5.8.

Since the flow is one-dimensional, $\mathbf{v} = v_z \mathbf{e}_z$. The equation of continuity, Eq. (A) in Table 1.1, simplifies to

$$\frac{\partial v_z}{\partial z} = 0 \quad \Rightarrow \quad v_z \neq v_z(z). \tag{5.5-1}$$

Therefore, $v_z = v_z(x, y)$. The components of the equation of motion, Eqs. (D), (E), and (F) in Table 1.5, simplify to

$$x\text{-component} \qquad 0 = \frac{\partial P}{\partial x} + \rho g, \tag{5.5-2}$$

$$y\text{-component} \qquad 0 = \frac{\partial P}{\partial y}, \tag{5.5-3}$$

$$z\text{-component} \qquad 0 = -\frac{\partial P}{\partial z} + \mu \left(\frac{\partial^2 v_z}{\partial x^2} + \frac{\partial^2 v_z}{\partial x^2} \right). \tag{5.5-4}$$

Defining the modified pressure as

$$\mathcal{P} = P + \rho g x. \tag{5.5-5}$$

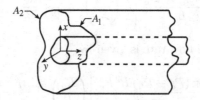

Figure 5.8   Flow in a straight duct of arbitrary cross-section.

reduces Eqs. (5.5-2)–(5.5-4) to

$$\frac{\partial P}{\partial x} = 0, \tag{5.5-6}$$

$$\frac{\partial P}{\partial y} = 0, \tag{5.5-7}$$

$$\frac{\partial P}{\partial z} = \mu \left(\frac{\partial^2 v_z}{\partial x^2} + \frac{\partial^2 v_z}{\partial x^2} \right). \tag{5.5-8}$$

Equations (5.5-6) and (5.5-7) imply that $P \neq P(x)$ and $P \neq P(y)$, respectively. Therefore, $P = P(z)$ and Eq. (5.5-8) takes the following form:

$$\underbrace{\frac{dP}{dz}}_{\text{function of } z} = \mu \underbrace{\left(\frac{\partial^2 v_z}{\partial x^2} + \frac{\partial^2 v_z}{\partial x^2} \right)}_{\text{function of } x \text{ and } y}. \tag{5.5-9}$$

While the left-hand side of Eq. (5.5-9) is a function of z only, the right-hand side is dependent on x and y. This is possible if and only if both sides of Eq. (5.5-9) are equal to a constant, say λ. Hence,

$$\frac{dP}{dz} = \lambda \quad \Rightarrow \quad \lambda = -\frac{P_o - P_L}{L}, \tag{5.5-10}$$

where P_o and P_L are the values of P at $z = 0$ and $z = L$, respectively. Substitution of Eq. (5.5-10) into Eq. (5.5-9) gives

$$\frac{\partial^2 v_z}{\partial x^2} + \frac{\partial^2 v_z}{\partial y^2} = -\frac{P_o - P_L}{\mu L}. \tag{5.5-11}$$

The no-slip boundary condition on the conduit surfaces is expressed as

$$v_z = 0 \qquad \text{on } A_1(x, y) \quad \text{and} \quad A_2(x, y). \tag{5.5-12}$$

Equation (5.5-11) is a nonhomogeneous partial differential equation. To make it homogeneous, let us use the transformation

$$v_z(x, y) = \varphi(x, y) - \frac{P_o - P_L}{4\mu L}(x^2 + y^2), \tag{5.5-13}$$

so that Eq. (5.5-11) reduces to a homogeneous partial differential equation of the form

$$\frac{\partial^2 \varphi}{\partial x^2} + \frac{\partial^2 \varphi}{\partial y^2} = 0. \tag{5.5-14}$$

In vector notation,[7] Eq. (5.5-14) is expressed as

$$\nabla^2 \varphi = 0. \tag{5.5-15}$$

The transformation given by Eq. (5.5-13) expresses the boundary conditions as

$$\varphi = \frac{(\mathcal{P}_o - \mathcal{P}_L)}{4\mu L}(x^2 + y^2) \qquad \text{on } A_1(x,y) \text{ and } A_2(x,y). \tag{5.5-16}$$

Once Eq. (5.5-15) subject to the boundary conditions defined by Eq. (5.5-16) is solved for φ, the solution for the axial velocity is obtained from Eq. (5.5-13).

5.5.1 *Flow through a concentric annulus*

With the help of Eq. (A.8-8) in Appendix A, Eq. (5.5-15) is expressed in the cylindrical coordinate system as

$$\nabla^2 \varphi = \frac{1}{r}\frac{d}{dr}\left(r\frac{d\varphi}{dr}\right) = 0 \qquad \Rightarrow \qquad \frac{d}{dr}\left(r\frac{d\varphi}{dr}\right) = 0. \tag{5.5-17}$$

The solution of Eq. (5.5-17) is given by

$$\varphi = C_1 \ln r + C_2. \tag{5.5-18}$$

The equations representing the surfaces of the inner and outer cylinders are

$$x^2 + y^2 = \kappa^2 R^2 \qquad \text{on } A_1, \tag{5.5-19}$$

$$x^2 + y^2 = R^2 \qquad \text{on } A_2. \tag{5.5-20}$$

Therefore, the boundary conditions, defined by Eq. (5.5-16), take the form

$$\text{at } r = \kappa R \qquad \varphi = \frac{(\mathcal{P}_o - \mathcal{P}_L)}{4\mu L}\kappa^2 R^2, \tag{5.5-21}$$

$$\text{at } r = R \qquad \varphi = \frac{(\mathcal{P}_o - \mathcal{P}_L)}{4\mu L}R^2. \tag{5.5-22}$$

The use of Eq. (5.5-22) in Eq. (5.5-18) gives

$$\frac{(\mathcal{P}_o - \mathcal{P}_L)R^2}{4\mu L} = C_1 \ln R + C_2 \quad \Rightarrow \quad C_2 = \frac{(\mathcal{P}_o - \mathcal{P}_L)R^2}{4\mu L} - C_1 \ln R. \tag{5.5-23}$$

---

[7]See Eq. (A.8-7) in Appendix A.

Substitution of Eq. (5.5-23) into Eq. (5.5-18) and rearrangement yield

$$\varphi = C_1 \ln\left(\frac{r}{R}\right) + \frac{(\mathcal{P}_o - \mathcal{P}_L)R^2}{4\mu L}. \tag{5.5-24}$$

Application of the boundary condition, defined by Eq. (5.5-21), results in

$$\frac{(\mathcal{P}_o - \mathcal{P}_L)\kappa^2 R^2}{4\mu L} = C_1 \ln\kappa + \frac{(\mathcal{P}_o - \mathcal{P}_L)R^2}{4\mu L}, \tag{5.5-25}$$

or

$$C_1 = -\frac{(\mathcal{P}_o - \mathcal{P}_L)R^2}{4\mu L}\frac{(1 - \kappa^2)}{\ln(1/\kappa)}. \tag{5.5-26}$$

Substitution of Eq. (5.5-26) into Eq. (5.5-24) leads to

$$\varphi = \frac{(\mathcal{P}_o - \mathcal{P}_L)R^2}{4\mu L}\left[1 - \frac{(1 - \kappa^2)}{\ln(1/\kappa)}\ln\left(\frac{r}{R}\right)\right]. \tag{5.5-27}$$

Thus, the velocity distribution is determined from Eq. (5.5-13) as

$$\begin{aligned} v_z &= \frac{(\mathcal{P}_o - \mathcal{P}_L)R^2}{4\mu L}\left[1 - \frac{(1 - \kappa^2)}{\ln(1/\kappa)}\ln\left(\frac{r}{R}\right)\right] - \frac{(\mathcal{P}_o - \mathcal{P}_L)r^2}{4\mu L} \\ &= \frac{(\mathcal{P}_o - \mathcal{P}_L)R^2}{4\mu L}\left[1 - \left(\frac{r}{R}\right)^2 - \frac{(1 - \kappa^2)}{\ln(1/\kappa)}\ln\left(\frac{r}{R}\right)\right], \end{aligned} \tag{5.5-28}$$

which is identical with Eq. (5.3-8).

5.5.2 *Flow in a rectangular duct*

Consider steady laminar flow of an incompressible Newtonian fluid in the z-direction through a rectangular duct as shown in Figure 5.9. The governing equation is given by Eq. (5.5-11), i.e.,

$$\frac{\partial^2 v_z}{\partial x^2} + \frac{\partial^2 v_z}{\partial y^2} = -\frac{\mathcal{P}_o - \mathcal{P}_L}{\mu L}. \tag{5.5-29}$$

The boundary conditions are given by

$$\begin{array}{lll} \text{at } x = 0 & v_z = 0, & \tag{5.5-30} \\ \text{at } x = b & v_z = 0, & \tag{5.5-31} \\ \text{at } y = 0 & v_z = 0, & \tag{5.5-32} \\ \text{at } y = a & v_z = 0. & \tag{5.5-33} \end{array}$$

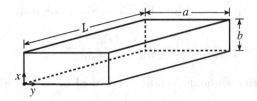

Figure 5.9   Flow in a rectangular duct.

The transformation

$$v_z(x,y) = \varphi(x,y) - \frac{(\mathcal{P}_o - \mathcal{P}_L)}{2\mu L}\, y(y-a).$$ (5.5-34)

reduces Eqs. (5.5-29)–(5.5-33) to the following form:

$$\frac{\partial^2 \varphi}{\partial x^2} + \frac{\partial^2 \varphi}{\partial y^2} = 0.$$ (5.5-35)

$$\text{at } x = 0 \qquad \varphi = \frac{(\mathcal{P}_o - \mathcal{P}_L)}{2\mu L}\, y(y-a),$$ (5.5-36)

$$\text{at } x = b \qquad \varphi = \frac{(\mathcal{P}_o - \mathcal{P}_L)}{2\mu L}\, y(y-a),$$ (5.5-37)

$$\text{at } y = 0 \qquad \varphi = 0,$$ (5.5-38)

$$\text{at } y = a \qquad \varphi = 0.$$ (5.5-39)

Since the partial differential equation and the boundary conditions in the y-direction are homogeneous,[8] the solution can be obtained by the method of separation of variables.

The separation of variables method assumes that the solution can be represented as a product of two functions of the form

$$\varphi(x,y) = F(x)\, G(y).$$ (5.5-40)

Substitution of Eq. (5.5-40) into Eq. (5.5-35) and rearrangement give

$$\frac{1}{F}\frac{d^2 F}{dx^2} = -\frac{1}{G}\frac{d^2 G}{dy^2}.$$ (5.5-41)

---

[8] A linear differential equation or a linear boundary condition is said to be homogeneous if, when satisfied by a function f, it is also satisfied by βf, where β is an arbitrary constant.

While the left-hand side of Eq. (5.5-41) is a function of x only, the right-hand side is dependent only on y. This is possible only if both sides of Eq. (5.5-41) are equal to a constant, say λ^2, i.e.,

$$\frac{1}{F}\frac{d^2 F}{dx^2} = -\frac{1}{G}\frac{d^2 G}{dy^2} = \lambda^2. \tag{5.5-42}$$

Equation (5.5-42) results in two ordinary differential equations. The equation for G is

$$\frac{d^2 G}{dy^2} + \lambda^2 G = 0, \tag{5.5-43}$$

and it is subject to the boundary conditions

$$\text{at } y = 0 \quad G = 0, \tag{5.5-44}$$

$$\text{at } y = a \quad G = 0. \tag{5.5-45}$$

Note that Eq. (5.5-43) is a Sturm–Liouville equation with a weight function of unity. The solution of Eq. (5.5-43) is given by

$$G(y) = C_1 \sin(\lambda y) + C_2 \cos(\lambda y), \tag{5.5-46}$$

where C_1 and C_2 are constants. The use of Eq. (5.5-44) gives $C_2 = 0$. Application of Eq. (5.5-45) gives

$$\sin(\lambda a) = 0 \quad \Rightarrow \quad \lambda_n a = n\pi \quad n = 1, 2, \ldots \tag{5.5-47}$$

For each eigenvalue λ_n, the corresponding eigenfunction is given by

$$G_n(y) = C_1 \sin\left(\frac{n\pi y}{a}\right). \tag{5.5-48}$$

On the other hand, the equation for F is

$$\frac{d^2 F_n}{dx^2} - \lambda_n^2 F_n = 0. \tag{5.5-49}$$

The solution of Eq. (5.5-49) is

$$F_n(x) = C_3 \sinh\left(\lambda_n x\right) + C_4 \cosh\left(\lambda_n x\right), \tag{5.5-50}$$

where C_3 and C_4 are constants. Substitution of Eqs. (5.5-48) and (5.5-50) into Eq. (5.5-40) gives

$$\varphi_n(x, y) = \sin\left(\frac{n\pi y}{a}\right)\left[A_n \sinh\left(\frac{n\pi x}{a}\right) + B_n \cosh\left(\frac{n\pi x}{a}\right)\right], \tag{5.5-51}$$

where $A_n = C_1 C_3$ and $B_n = C_1 C_4$.

If φ_1 and φ_2 are the solutions satisfying the partial differential equations and the boundary conditions, then the linear combination of the solutions, i.e., $\alpha_1 \varphi_1 + \alpha_2 \varphi_2$, also satisfies the partial differential equation and the boundary conditions. Therefore, the complete solution is

$$\varphi = \sum_{n=1}^{\infty} \sin\left(\frac{n\pi y}{a}\right) \left[A_n \sinh\left(\frac{n\pi x}{a}\right) + B_n \cosh\left(\frac{n\pi x}{a}\right)\right]. \tag{5.5-52}$$

Application of the boundary condition defined by Eq. (5.5-36) gives

$$\frac{(\mathcal{P}_o - \mathcal{P}_L)}{2\mu L} y(y - a) = \sum_{n=1}^{\infty} B_n \sin\left(\frac{n\pi y}{a}\right). \tag{5.5-53}$$

Since the eigenfunctions are simply orthogonal, multiplication of Eq. (5.5-53) by $\sin(m\pi y/a)\, dy$ and integration from $y = 0$ to $y = a$ give

$$\frac{(\mathcal{P}_o - \mathcal{P}_L)}{2\mu L} \int_0^a (y^2 - ay) \sin\left(\frac{m\pi y}{a}\right) dy$$
$$= \sum_{n=1}^{\infty} B_n \int_0^a \sin\left(\frac{m\pi y}{a}\right) \sin\left(\frac{n\pi y}{a}\right) dy. \tag{5.5-54}$$

The integral on the right-hand side of Eq. (5.5-54) is zero when $n \neq m$ and non-zero when $n = m$. Therefore, the summation drops out when $n = m$ and Eq. (5.5-54) reduces to the form

$$\frac{(\mathcal{P}_o - \mathcal{P}_L)}{2\mu L} \int_0^a (y^2 - ay) \sin\left(\frac{n\pi y}{a}\right) dy = B_n \int_0^a \sin^2\left(\frac{n\pi y}{a}\right) dy. \tag{5.5-55}$$

Evaluation of the integrals leads to

$$B_n = \left[(-1)^n - 1\right] \frac{(\mathcal{P}_o - \mathcal{P}_L)a^2}{\mu L (n\pi)^3}. \tag{5.5-56}$$

Application of the boundary condition defined by Eq. (5.5-37) gives

$$\frac{(\mathcal{P}_o - \mathcal{P}_L)}{2\mu L} y(y - a)$$
$$= \sum_{n=1}^{\infty} \sin\left(\frac{n\pi y}{a}\right) \left[A_n \sinh\left(\frac{n\pi b}{a}\right) + B_n \cosh\left(\frac{n\pi b}{a}\right)\right]. \tag{5.5-57}$$

The similar procedure used in the evaluation of B_n leads to

$$\frac{(\mathcal{P}_o - \mathcal{P}_L)}{2\mu L} \int_0^a (y^2 - ay) \sin\left(\frac{n\pi y}{a}\right) dy$$

$$= A_n \sinh\left(\frac{n\pi b}{a}\right) \int_0^a \sin^2\left(\frac{n\pi y}{a}\right) dy + B_n \cosh\left(\frac{n\pi b}{a}\right) \int_0^a \sin^2\left(\frac{n\pi y}{a}\right) dy. \tag{5.5-58}$$

Evaluation of the integrals gives

$$A_n = \left[\frac{1 - \cosh(n\pi b/a)}{\sinh(n\pi b/a)}\right] B_n. \tag{5.5-59}$$

Finally, the velocity distribution is obtained by substituting Eq. (5.5-52) into Eq. (5.5-34). The result is

$$v_z = \sum_{n=1}^{\infty} \sin\left(\frac{n\pi y}{a}\right) \left[A_n \sinh\left(\frac{n\pi x}{a}\right) + B_n \cosh\left(\frac{n\pi x}{a}\right)\right]$$

$$- \frac{(\mathcal{P}_o - \mathcal{P}_L)}{2\mu L} y(y - a), \tag{5.5-60}$$

where A_n and B_n are defined by Eqs. (5.5-59) and (5.5-56), respectively.

The volumetric flow rate is obtained by integrating the velocity distribution over the cross-sectional area, i.e.,

$$\mathcal{Q} = \int_0^a \int_0^b v_z \, dx \, dy. \tag{5.5-61}$$

Substitution of Eq. (5.5-60) into Eq. (5.5-61) and integration lead to

$$\mathcal{Q} = \frac{(\mathcal{P}_o - \mathcal{P}_L)(a^2 + b^2)ab}{24\mu L} - \frac{8(\mathcal{P}_o - \mathcal{P}_L)}{\mu L \pi^5} \sum_{n=1}^{\infty} \frac{1}{(2n-1)^5}$$

$$\times \left\{ a^4 \tanh\left[\frac{(2n-1)\pi b}{2a}\right] + b^4 \tanh\left[\frac{(2n-1)\pi a}{2b}\right] \right\}. \tag{5.5-62}$$

Chapter 6

Flow Due to Pressure Gradient and Motion of Boundaries

This chapter deals with fluid flow resulting from the combination of pressure gradient and motion of boundaries. The problems presented in this chapter will be analyzed with the following assumptions:

- steady-state,
- incompressible Newtonian fluid,
- fully developed laminar flow,
- constant physical properties.

The imposed pressure gradient may or may not be in the direction of the movement of a solid boundary enclosing the fluid. When the moving boundary and the pressure gradient are in the same or completely opposite direction, the resulting flow is often referred to as *Generalized Couette Flow*.

6.1 Generalized Couette Flow Between Parallel Plates

Consider one-dimensional flow of a Newtonian fluid between two parallel plates as shown in Figure 6.1. The plate at $x = H$ is fixed, whereas the one at $x = 0$ moves in the positive z-direction with a uniform velocity V. A pressure gradient is also imposed in the z-direction.

Simplification of the components of the equation of motion follows the same lines as explained in Section 5.1. Thus, the governing equation of motion for velocity is given by Eq. (5.1-11), i.e.,

$$-\mu \frac{d^2 v_z}{dx^2} = \frac{\mathcal{P}_o - \mathcal{P}_L}{L}. \tag{6.1-1}$$

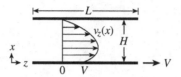

Figure 6.1   Generalized Couette flow between parallel plates.

Integration of Eq. (6.1-1) twice results in

$$v_z = -\frac{\mathcal{P}_o - \mathcal{P}_L}{2\,\mu L}\,x^2 + C_1\,x + C_2, \qquad (6.1\text{-}2)$$

where C_1 and C_2 are integration constants. The use of the boundary conditions

$$\text{at } x = 0 \qquad v_z = V, \qquad (6.1\text{-}3\text{a})$$

$$\text{at } x = H \qquad v_z = 0, \qquad (6.1\text{-}3\text{b})$$

gives the velocity distribution as

$$v_z = \underbrace{\frac{(\mathcal{P}_o - \mathcal{P}_L)\,H^2}{2\,\mu L}\left[\frac{x}{H} - \left(\frac{x}{H}\right)^2\right]}_{A} + \underbrace{V\left(1 - \frac{x}{H}\right)}_{B}. \qquad (6.1\text{-}4)$$

Equation (6.1-4) is the superposition of the solution of two problems[1]: the term A is the velocity distribution for flow between two fixed parallel plates due to a pressure gradient, Eq. (5.1-14); the term B is the velocity distribution for flow between two parallel plates, one of which ($x = 0$) is moving, with no pressure gradient, Eq. (4.1-9). A representative velocity distribution is also shown in Figure 6.1.

From Table 1.2, the only non-zero shear stress component is

$$\tau_{xz} = -\mu\frac{dv_z}{dx}. \qquad (6.1\text{-}5)$$

Differentiation of Eq. (6.1-4) with respect to x and substitution of the resulting equation into Eq. (6.1-5) give the shear stress distribution as

$$\tau_{xz} = \frac{(\mathcal{P}_o - \mathcal{P}_L)\,H}{2L}\left[2\left(\frac{x}{H}\right) - 1\right] + \frac{\mu V}{H}. \qquad (6.1\text{-}6)$$

---

[1]This property is valid only for Newtonian fluids in which the relation between shear stress and shear rate is linear.

The volumetric flow rate can be determined by integrating the velocity distribution over the cross-sectional area, i.e.,

$$Q = \int_0^W \int_0^H v_z \, dx \, dy = W \int_0^H v_z \, dx, \tag{6.1-7}$$

where W is the width of the plate. Substitution of Eq. (6.1-4) into Eq. (6.1-7) and integration result in

$$Q = \underbrace{\frac{(\mathcal{P}_o - \mathcal{P}_L) W H^3}{12 \, \mu L}}_{\text{Eq. (5.1-18)}} + \underbrace{\frac{HWV}{2}}_{\text{Eq. (4.1-11)}}. \tag{6.1-8}$$

Dividing the volumetric flow rate by the flow area gives the average velocity as

$$\langle v_z \rangle = \frac{Q}{WH} = \frac{(\mathcal{P}_o - \mathcal{P}_L) H^2}{12 \, \mu L} + \frac{V}{2}. \tag{6.1-9}$$

6.1.1 Calculation of drag force

The drag force exerted by the fluid on the plate surfaces is given by Eq. (5.1-21), i.e.,

$$F_D = \int_0^L \int_0^W - \tau_{xz}|_{x=0} \, dy \, dz + \int_0^L \int_0^W \tau_{xz}|_{x=H} \, dy \, dz. \tag{6.1-10}$$

The use of Eq. (6.1-6) in Eq. (6.1-10) leads to

$$F_D = (\mathcal{P}_o - \mathcal{P}_L) W H. \tag{6.1-11}$$

The power required to pump the fluid is calculated with the help of Eq. (1.2-26) as

$$\dot{W} = F_D \langle v_z \rangle = (\mathcal{P}_o - \mathcal{P}_L) W H \langle v_z \rangle = (\mathcal{P}_o - \mathcal{P}_L) Q, \tag{6.1-12}$$

which is the product of pressure drop and volumetric flow rate.

6.1.2 Effect of using a different coordinate system

Let us rework the problem using the coordinate system shown in Figure 6.2. As shown in Section 5.1.3, the volumetric flow rate expression, Eq. (6.1-8), does not depend on the coordinate system. If we replace H in Eq. (6.1-4) by $x + (H/2)$, the velocity distribution becomes

$$v_z = \frac{(\mathcal{P}_o - \mathcal{P}_L) H^2}{8 \mu L} \left[1 - 4 \left(\frac{x}{H} \right)^2 \right] + V \left(\frac{1}{2} - \frac{x}{H} \right). \tag{6.1-13}$$

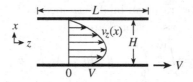

Figure 6.2   Generalized Couette flow between parallel plates.

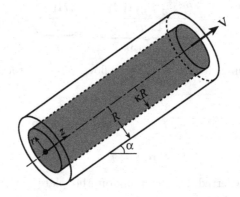

Figure 6.3   Generalized Couette flow in an annulus.

6.2   Generalized Couette Flow in a Concentric Annulus

Consider fluid flow through an inclined annulus under the action of pressure gradient together with the movement of the inner cylinder with a constant velocity V as shown in Figure 6.3. The radii of the inner and outer cylinders are κR and R, respectively.

The development of the pressure and velocity distributions follows the same lines for flow in a circular tube given in Section 5.2. Therefore,

$$\mathcal{P} - \mathcal{P}_o = -\left(\frac{\mathcal{P}_o - \mathcal{P}_L}{L}\right) z + \rho g \cos \alpha (r \sin \theta + R), \qquad (6.2\text{-}1)$$

and

$$v_z = -\frac{(\mathcal{P}_o - \mathcal{P}_L)}{4\mu L} r^2 + C_1 \ln r + C_2. \qquad (6.2\text{-}2)$$

Application of the boundary condition on the outer tube surface, i.e.,

$$\text{at } r = R \qquad v_z = 0, \qquad (6.2\text{-}3)$$

to Eq. (6.2-2) yields

$$C_2 = \frac{(\mathcal{P}_o - \mathcal{P}_L)\,R^2}{4\mu L} - C_1 \ln R.$$

(6.2-4)

Substitution of Eq. (6.2-4) into Eq. (6.2-2) and rearrangement give

$$v_z = \frac{(\mathcal{P}_o - \mathcal{P}_L)\,R^2}{4\mu L}\left[1 - \left(\frac{r}{R}\right)^2\right] + C_1 \ln\left(\frac{r}{R}\right).$$

(6.2-5)

Application of the boundary condition on the inner tube surface, i.e.,

$$\text{at } r = \kappa R \qquad v_z = V,$$

(6.2-6)

to Eq. (6.2-5) results in

$$C_1 = \frac{(\mathcal{P}_o - \mathcal{P}_L)\,R^2}{4\mu L}\frac{(1 - \kappa^2)}{\ln(1/\kappa)} - \frac{V}{\ln(1/\kappa)}.$$

(6.2-7)

Substitution of Eq. (6.2-7) into Eq. (6.2-5) leads to the following velocity distribution:

$$v_z = \frac{(\mathcal{P}_o - \mathcal{P}_L)\,R^2}{4\mu L}\left[1 - \left(\frac{r}{R}\right)^2 + \frac{(1 - \kappa^2)}{\ln(1/\kappa)}\ln\left(\frac{r}{R}\right)\right]$$
$$- \frac{V}{\ln(1/\kappa)}\ln\left(\frac{r}{R}\right).$$

(6.2-8)

From Table 1.3, the only non-zero shear stress component is

$$\tau_{rz} = -\mu\frac{dv_z}{dr}.$$

(6.2-9)

Differentiation of Eq. (6.2-8) with respect to r and substitution of the resulting equation into Eq. (6.2-9) give the shear stress distribution as

$$\tau_{rz} = \frac{(\mathcal{P}_o - \mathcal{P}_L)R}{2L}\left[\frac{r}{R} - \frac{(1 - \kappa^2)}{2\ln(1/\kappa)}\frac{R}{r}\right] + \frac{\mu V}{\ln(1/\kappa)}\frac{1}{r}.$$

(6.2-10)

The volumetric flow rate can be determined by integrating the velocity distribution over the annular cross-sectional area, i.e.,

$$Q = \int_0^{2\pi}\int_{\kappa R}^{R} v_z r\,dr\,d\theta = 2\pi\int_{\kappa R}^{R} v_z r\,dr.$$

(6.2-11)

Before substituting Eq. (6.2-8) into Eq. (6.2-11), introduction of the dimensionless distance ξ, defined by

$$\xi = \frac{r}{R},$$

(6.2-12)

simplifies the resulting expression as

$$Q = \frac{(\mathcal{P}_o - \mathcal{P}_L)\pi R^4}{2\mu L} \int_\kappa^1 \left[1 - \xi^2 + \frac{(1 - \kappa^2)}{\ln(1/\kappa)} \ln \xi \right] \xi \, d\xi$$
$$- \frac{2\pi R^2 V}{\ln(1/\kappa)} \int_\kappa^1 \xi \ln \xi \, d\xi. \tag{6.2-13}$$

Integration of Eq. (6.2-13) leads to

$$Q = \frac{\pi (\mathcal{P}_o - \mathcal{P}_L) R^4}{8\mu L} \left[1 - \kappa^4 - \frac{(1 - \kappa^2)^2}{\ln(1/\kappa)} \right]$$
$$+ \frac{\pi R^2 V}{2 \ln(1/\kappa)} \left[1 - \kappa^2 - 2\kappa^2 \ln(1/\kappa) \right]. \tag{6.2-14}$$

When $V = 0$, Eqs. (6.2-8) and (6.2-14) reduce to Eqs. (5.3-8) and (5.3-14), respectively.

6.2.1 *Calculation of drag force*

The drag force exerted by the fluid on the inner moving cylinder is given by Eq. (4.1-15), in which $\lambda = \mathbf{e}_z$, and the unit normal vector directed from the fluid to the tube surface is $\mathbf{n} = -\mathbf{e}_r$. Therefore, Eq. (4.1-15) simplifies to

$$F_D = \int_0^L \int_0^{2\pi} - \tau_{rz}|_{r=\kappa R} \, \kappa R \, d\theta \, dz. \tag{6.2-15}$$

The use of Eq. (6.2-10) in Eq. (6.2-15) leads to

$$F_D = (\mathcal{P}_o - \mathcal{P}_L) \pi R^2 (1 - \kappa^2). \tag{6.2-16}$$

6.3 Helical Flow Through a Concentric Annulus

Consider steady flow of a Newtonian fluid in a concentric annulus when there is both an axial and tangential flow as shown in Figure 6.4. While the axial flow is due to the imposed pressure gradient, the tangential flow arises from rotation of the inner cylinder at a constant angular velocity Ω.

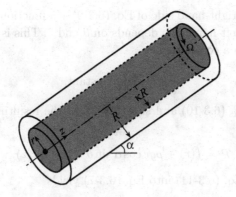

Figure 6.4   Helical flow in a concentric annulus.

Postulating $v_r = 0$, $v_\theta = v_\theta(r)$, and $v_z = v_z(r)$, the components of the equation of motion, Eqs. (D), (E), and (F) in Table 1.6, simplify to

r-component $\quad -\rho \dfrac{v_\theta^2}{r} = -\dfrac{\partial P}{\partial r} + \rho g \cos\alpha \sin\theta,$ $\hspace{2cm}$ (6.3-1)

θ-component $\quad 0 = -\dfrac{1}{r}\dfrac{\partial P}{\partial \theta} + \mu \left\{ \dfrac{d}{dr}\left[\dfrac{1}{r}\dfrac{d}{dr}(r v_\theta) \right] \right\}$

$\hspace{4cm} + \rho g \cos\alpha \cos\theta,$ $\hspace{2cm}$ (6.3-2)

z-component $\quad 0 = -\dfrac{\partial P}{\partial z} + \dfrac{\mu}{r}\dfrac{d}{dr}\left(r \dfrac{dv_z}{dr} \right) - \rho g \sin\alpha.$ $\hspace{1cm}$ (6.3-3)

Defining the modified pressure as

$$\mathcal{P} = P + \rho g z \sin\alpha,$$ $\hspace{2cm}$ (6.3-4)

reduces Eqs. (6.3-1)–(6.3-3) to

$$\frac{\partial \mathcal{P}}{\partial r} = \rho \frac{v_\theta^2}{r} + \rho g \cos\alpha \sin\theta,$$ $\hspace{2cm}$ (6.3-5)

$$\frac{\partial \mathcal{P}}{\partial \theta} = \mu r \frac{d}{dr}\left[\frac{1}{r}\frac{d}{dr}(r v_\theta) \right] + \rho g r \cos\alpha \cos\theta,$$ $\hspace{1cm}$ (6.3-6)

$$\frac{\partial \mathcal{P}}{\partial z} = \frac{\mu}{r}\frac{d}{dr}\left(r \frac{dv_z}{dr} \right).$$ $\hspace{2cm}$ (6.3-7)

Integration of Eq. (6.3-5) with respect to r gives

$$\mathcal{P} = f(r) + \rho g r \cos\alpha \sin\theta + h(\theta, z).$$ $\hspace{1cm}$ (6.3-8)

Substitution of Eq. (6.3-8) into Eq. (6.3-6) yields

$$\underbrace{\frac{\partial h}{\partial \theta}}_{\text{function of } \theta \text{ and } z} = \underbrace{\mu r \frac{d}{dr}\left[\frac{1}{r}\frac{d}{dr}(r v_\theta) \right]}_{\text{function of } r}.$$ $\hspace{1cm}$ (6.3-9)

The term on the right-hand side of Eq. (6.3-9) is a function only of r, while the term on the left-hand side depends on θ and z. This is possible if both sides are equal to a constant, i.e.,

$$\frac{\partial h}{\partial \theta} = A = \text{constant}. \tag{6.3-10}$$

Integration of Eq. (6.3-10) and substitution of the resulting equation into Eq. (6.3-8) give

$$\mathcal{P} = f(r) + \rho g r \cos \alpha \sin \theta + A\theta + l(z). \tag{6.3-11}$$

Substitution of Eq. (6.3-11) into Eq. (6.3-7) yields

$$\underbrace{\frac{dl}{dz}}_{\text{function of } z} = \underbrace{\frac{\mu}{r} \frac{d}{dr} \left(r \frac{dv_z}{dr} \right)}_{\text{function of } r}. \tag{6.3-12}$$

The term on the right-hand side of Eq. (6.3-12) is a function only of r, while the term on the left-hand side depends only on z. This is possible if both sides are equal to a constant, i.e.,

$$\frac{dl}{dz} = B = \text{constant}. \tag{6.3-13}$$

Integration of Eq. (6.3-13) and substitution of the resulting equation into Eq. (6.3-11) give

$$\mathcal{P} = f(r) + \rho g r \cos \alpha \sin \theta + A\theta + B z + C, \tag{6.3-14}$$

where C is a constant of integration. Since

$$\mathcal{P}(r, \theta, z) = \mathcal{P}(r, \theta + 2\pi, z), \tag{6.3-15}$$

then the constant A in Eq. (6.3-14) must be zero. Therefore, the pressure distribution becomes

$$\mathcal{P} = f(r) + \rho g r \cos \alpha \sin \theta + B z + C. \tag{6.3-16}$$

Differentiation of Eq. (6.3-16) with respect to z yields

$$\frac{\partial \mathcal{P}}{\partial z} = B = -\frac{\mathcal{P}_o - \mathcal{P}_L}{L}, \tag{6.3-17}$$

where $\mathcal{P}_o$ and $\mathcal{P}_L$ are the values of $\mathcal{P}$ at $z = 0$ and $z = L$, respectively. Substitution of Eq. (6.3-17) into Eq. (6.3-7) gives

$$\frac{d}{dr} \left(r \frac{dv_z}{dr} \right) = -\frac{(\mathcal{P}_o - \mathcal{P}_L)}{\mu L} r. \tag{6.3-18}$$

Integration of Eq. (6.3-18) twice results in

$$v_z = -\frac{(\mathcal{P}_o - \mathcal{P}_L)}{4\,\mu L}\,r^2 + C_1 \ln r + C_2, \tag{6.3-19}$$

where C_1 and C_2 are constants of integration. The use of the boundary conditions

$$\text{at } r = \kappa R \qquad v_z = 0, \tag{6.3-20a}$$

$$\text{at } r = R \qquad v_z = 0, \tag{6.3-20b}$$

gives the velocity distribution as

$$v_z = \frac{(\mathcal{P}_o - \mathcal{P}_L)\,R^2}{4\mu L}\left[1 - \left(\frac{r}{R}\right)^2 + \frac{(1-\kappa^2)}{\ln(1/\kappa)}\ln\left(\frac{r}{R}\right)\right]. \tag{6.3-21}$$

Substitution of Eq. (6.3-8) into Eq. (6.3-6) leads to

$$\frac{d}{dr}\left[\frac{1}{r}\frac{d}{dr}(rv_\theta)\right] = 0. \tag{6.3-22}$$

Integration of Eq. (6.3-22) twice gives

$$v_\theta = C_3\,r + \frac{C_4}{r}, \tag{6.3-23}$$

where C_3 and C_4 are constants of integration. The use of the boundary conditions

$$\text{at } r = \kappa R \qquad v_\theta = \kappa R\Omega, \tag{6.3-24a}$$

$$\text{at } r = R \qquad v_\theta = 0, \tag{6.3-24b}$$

gives the velocity distribution as

$$v_\theta = \frac{\kappa R\Omega}{(1/\kappa) - \kappa}\left(\frac{R}{r} - \frac{r}{R}\right). \tag{6.3-25}$$

From Table 1.3, the non-zero shear stress components are

$$\tau_{rz} = \tau_{zr} = -\mu\frac{dv_z}{dr}, \tag{6.3-26}$$

$$\tau_{r\theta} = \tau_{\theta r} = -\mu r\frac{d}{dr}\left(\frac{v_\theta}{r}\right). \tag{6.3-27}$$

The use of Eq. (6.3-21) in Eq. (6.3-26) gives

$$\tau_{rz} = \frac{(\mathcal{P}_o - \mathcal{P}_L)R}{2L}\left[\frac{r}{R} - \frac{(1-\kappa^2)}{2\ln(1/\kappa)}\frac{R}{r}\right], \tag{6.3-28}$$

which is identical with Eq. (5.3-10). On the other hand, the use of Eq. (6.3-25) in Eq. (6.3-27) results in

$$\tau_{r\theta} = \frac{2\mu\kappa^2 R^2 \Omega}{1 - \kappa^2} \frac{1}{r^2}. \tag{6.3-29}$$

The volumetric flow rate can be determined by integrating the velocity distribution over the annular cross-sectional area, i.e.,

$$Q = \int_0^{2\pi} \int_{\kappa R}^R v_z r \, dr \, d\theta = 2\pi \int_{\kappa R}^R v_z r \, dr = 2\pi R^2 \int_\kappa^1 v_z \xi \, d\xi. \tag{6.3-30}$$

Substitution of Eq. (6.3-21) into Eq. (6.3-30) and integration leads to

$$Q = \frac{\pi (\mathcal{P}_o - \mathcal{P}_L) R^4}{8\mu L} \left[1 - \kappa^4 - \frac{(1 - \kappa^2)^2}{\ln(1/\kappa)} \right]. \tag{6.3-31}$$

6.3.1 *Calculation of torque*

The torque exerted by the fluid on the inner rotating cylinder is given by Eq. (4.2-29), in which the unit normal vector directed from the fluid to the inner tube surface is $\mathbf{n} = -\mathbf{e}_r$. Therefore, Eq. (4.2-29) becomes

$$T_z = -\int_0^L \int_0^{2\pi} \mathbf{e}_z \cdot \left[(r\mathbf{e}_r + z\mathbf{e}_z) \times (P\mathbf{e}_r + \mathbf{e}_r \cdot \tau_{ij}\mathbf{e}_i\mathbf{e}_j) \right] \kappa R \, d\theta \, dz. \tag{6.3-32}$$

Equation (6.3-32) simplifies to

$$T_z = -\int_0^L \int_0^{2\pi} (r\tau_{r\theta})|_{r=\kappa R} \, \kappa R \, d\theta \, dz. \tag{6.3-33}$$

Substitution of Eq. (6.3-29) into Eq. (6.3-33) and integration yield

$$T_z = \frac{4\pi\mu L(\kappa R)^2 \Omega}{\kappa^2 - 1}. \tag{6.3-34}$$

Chapter 7

Flow of Non-Newtonian Fluids

7.1 What is a Non-Newtonian Fluid?

The equation of motion, given by Eq. (1.2-17), i.e.,

$$\rho \frac{D\mathbf{v}}{Dt} = -\nabla P - \nabla \cdot \boldsymbol{\tau} + \rho \mathbf{g}, \tag{7.1-1}$$

holds for any fluid and for any coordinate system.

For an incompressible Newtonian fluid, the shear stress tensor is given by Eq. (1.2-19), i.e.,

$$\boldsymbol{\tau} = -\mu \dot{\boldsymbol{\gamma}}, \tag{7.1-2}$$

where μ is the viscosity and $\dot{\boldsymbol{\gamma}}$ is the symmetric rate of deformation tensor, defined by Eq. (1.2-20).

For an incompressible non-Newtonian fluid, the relationship between shear stress and rate of deformation tensor is expressed in the form

$$\boldsymbol{\tau} = -\eta \dot{\boldsymbol{\gamma}}, \tag{7.1-3}$$

where η is the non-Newtonian viscosity. Besides temperature, pressure, and composition, η is a function of the magnitudes of the rate of deformation or shear stress tensors. The magnitude of the rate of deformation tensor, also called the *shear rate*, is defined by

$$\dot{\gamma} = \left| \sqrt{\frac{1}{2} (\dot{\boldsymbol{\gamma}} : \dot{\boldsymbol{\gamma}})} \right|. \tag{7.1-4}$$

The magnitude of the shear stress tensor is

$$\tau = \left| \sqrt{\frac{1}{2} (\boldsymbol{\tau} : \boldsymbol{\tau})} \right|. \tag{7.1-5}$$

107

7.1.1 *Non-Newtonian fluid models*

The classification of non-Newtonian fluids is shown in Figure 7.1. Non-Newtonian fluids are broadly classified as being inelastic and viscoelastic.[1] For time-independent non-Newtonian fluids, shear stress is independent of time or duration of shear. On the other hand, shear stress is dependent on time or duration of shear for time-dependent non-Newtonian fluids.

While some time-independent non-Newtonian fluids exhibit a yield stress, i.e., the shear stress must exceed a certain value before the fluid deforms and flows,[2] the others do not have a yield stress. Time-independent non-Newtonian fluids that do not have a yield stress are classified as pseudoplastic and dilatant. In the case of pseudoplastic fluids, non-Newtonian viscosity decreases with increasing shear rate. Dilatant fluids are just the opposite of pseudoplastic fluids in that the non-Newtonian viscosity increases with increasing shear rate.

A typical example of a pseudoplastic fluid is the ink in an ordinary ballpoint pen. When the pen is not in use, the ink is so viscous that it does not flow. When one starts scribbling, the small ball on its point rolls. The rotation of the ball creates a shearing movement and lowers the non-Newtonian viscosity of the ink such that it flows on the paper. As soon as scribbling is stopped, the ink ceases to flow because the ball is no longer turning (provided that it is a good quality pen!).

Quicksand is a good example of a dilatant fluid. Often, the immediate reaction of a person who walks into quicksand is to attempt to escape as soon as possible. Yet, these movements create shearing. The more frightened a person becomes and the more he/she tries to escape, the greater the force

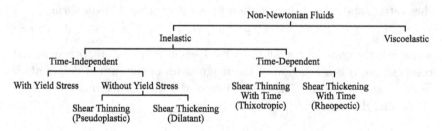

Figure 7.1   Classification of non-Newtonian fluids.

---

[1] A viscoelastic fluid has both viscous and elastic properties. Once the applied stress is released, a viscoelastic fluid either fully or partially returns to its original state.
[2] For example, in order to get a toothpaste to flow, the tube should be gently squeezed.

of his/her movement, which causes a corresponding increase in the rate of shear. In other words, the greater his/her efforts, the more viscous the quicksand becomes and the more difficult it is to escape.

Time-dependent non-Newtonian fluids are classified as thixotropic and rheopectic. A material whose non-Newtonian viscosity decreases with the time of shear is said to be thixotropic and one whose non-Newtonian viscosity increases with time of shear is said to be rheopectic. Thixotropic behavior is much more common than rheopectic behavior. Examples of thixotropic fluids are bentonite–water suspensions, gels, cosmetics, and nail polishes.

Shear stress–shear rate relationships for some of the non-Newtonian models are given as follows.

7.1.1.1 *Power-law fluid model*

The two-parameter *power-law* expression is given by

$$\eta = m\dot{\gamma}^{n-1}, \tag{7.1-6}$$

where m (in $N \cdot s^n/m^2$) and the dimensionless quantity n are constants characteristic of a polymeric fluid. When $m = \mu$ and $n = 1$, one obtains the Newtonian fluid. For pseudoplastic fluids $n < 1$, while $n > 1$ for dilatant fluids. Most polymeric fluids are pseudoplastic with n typically varying between 0.15 and 0.6. A power-law fluid reduces to a Newtonian fluid when $m = \mu$ and $n = 1$.

7.1.1.2 *Bingham fluid model*

A Bingham fluid exhibits a yield stress and the non-Newtonian viscosity is expressed as

$$\eta = \infty \qquad \text{when } \tau \leq \tau_o, \tag{7.1-7a}$$

$$\eta = \mu_o + \frac{\tau_o}{\dot{\gamma}} \qquad \text{when } \tau \geq \tau_o. \tag{7.1-7b}$$

The term τ_o is called the *yield stress*,[3] having the dimensions of force per unit area. The term μ_o, plastic viscosity, has the dimensions of viscosity. The Bingham model is primarily used for slurries and pastes.

---

[3]The yield stress is the stress at which a fluid just starts or stops to move. A Bingham fluid behaves as a solid at a stress below τ_o.

7.1.1.3   *Ellis fluid model*

For an Ellis fluid, the non-Newtonian viscosity is expressed as

$$\frac{\eta_o}{\eta} = 1 + \left(\frac{\tau}{\tau_{1/2}}\right)^{\alpha-1}, \tag{7.1-8}$$

where η_o is the value of the viscosity at zero shear, $\tau_{1/2}$ is the value of the shear stress at which $\eta = \eta_o/2$, and α is the Ellis fluid constant. An Ellis fluid reduces to a Newtonian fluid when $\eta_o = \mu$ and $\alpha = 1$.

7.1.1.4   *Herschel–Bulkley fluid model*

The three-constant Herschel–Bulkley fluid model is a generalization of the Bingham fluid model in which the shear stress expression is given by

$$\eta = \infty \qquad\qquad \text{when } \tau \leq \tau_o, \tag{7.1-9a}$$

$$\eta = m\dot\gamma^{n-1} + \frac{\tau_o}{\dot\gamma} \qquad \text{when } \tau \geq \tau_0. \tag{7.1-9b}$$

This model reduces to the power-law model when $\tau_o = 0$. On the other hand, it reduces to the Bingham model when $n = 1$ and $m = \mu_o$.

7.1.2   The procedure for solving flow problems involving non-Newtonian fluids

(1) Postulate the functional forms of the velocity components.

(2) Use Tables 1.2–1.4 for Newtonian fluids to determine the non-zero shear stress components. Replace μ in the Newtonian expression by η.

(3) Using the equation of motion in terms of τ given in Tables 1.5–1.7, determine the shear stress distribution. Note that this expression is valid for both Newtonian and non-Newtonian fluids.

(4) Insert the relation for $(\dot\gamma : \dot\gamma)$ or $(\tau : \tau)$ in the proper coordinates for the non-Newtonian fluid in question to find out η in the shear stress expression. The term $\frac{1}{2}(\dot\gamma : \dot\gamma)$ in the Cartesian, cylindrical, and spherical coordinate systems is given in Table 7.1.

(5) Equate the expressions obtained in steps (3) and (4) above and integrate the resulting equation to obtain the velocity distribution.

Table 7.1   The term $\frac{1}{2}(\dot{\gamma} : \dot{\gamma})$ in different coordinate systems.

Cartesian

$$\frac{1}{2}(\dot{\gamma} : \dot{\gamma}) = 2\left[\left(\frac{\partial v_x}{\partial x}\right)^2 + \left(\frac{\partial v_y}{\partial y}\right)^2 + \left(\frac{\partial v_z}{\partial z}\right)^2\right]$$

$$\times \left(\frac{\partial v_y}{\partial x} + \frac{\partial v_x}{\partial y}\right)^2 + \left(\frac{\partial v_z}{\partial y} + \frac{\partial v_y}{\partial z}\right)^2 + \left(\frac{\partial v_x}{\partial z} + \frac{\partial v_z}{\partial x}\right)^2 \qquad \text{(A)}$$

Cylindrical

$$\frac{1}{2}(\dot{\gamma} : \dot{\gamma}) = 2\left[\left(\frac{\partial v_r}{\partial r}\right)^2 + \left(\frac{1}{r}\frac{\partial v_\theta}{\partial \theta} + \frac{v_r}{r}\right)^2 + \left(\frac{\partial v_z}{\partial z}\right)^2\right]$$

$$+ \left[r\frac{\partial}{\partial r}\left(\frac{v_\theta}{r}\right) + \frac{1}{r}\frac{\partial v_r}{\partial \theta}\right]^2 + \left(\frac{1}{r}\frac{\partial v_z}{\partial \theta} + \frac{\partial v_\theta}{\partial z}\right)^2 + \left(\frac{\partial v_r}{\partial z} + \frac{\partial v_z}{\partial r}\right)^2 \qquad \text{(B)}$$

Spherical

$$\frac{1}{2}(\dot{\gamma} : \dot{\gamma}) = 2\left[\left(\frac{\partial v_r}{\partial r}\right)^2 + \left(\frac{1}{r}\frac{\partial v_\theta}{\partial \theta} + \frac{v_r}{r}\right)^2\right.$$

$$+ \left(\frac{1}{r\sin\theta}\frac{\partial v_\phi}{\partial \phi} + \frac{v_r}{r} + \frac{v_\theta \cot\theta}{r}\right)^2\bigg] + \left[r\frac{\partial}{\partial r}\left(\frac{v_\theta}{r}\right) + \frac{1}{r}\frac{\partial v_r}{\partial \theta}\right]^2$$

$$+ \left[\frac{\sin\theta}{r}\frac{\partial}{\partial \theta}\left(\frac{v_\phi}{\sin\theta}\right) + \frac{1}{r\sin\theta}\frac{\partial v_\theta}{\partial \phi}\right]^2 + \left[\frac{1}{r\sin\theta}\frac{\partial v_r}{\partial \phi} + r\frac{\partial}{\partial r}\left(\frac{v_\phi}{r}\right)\right]^2 \qquad \text{(C)}$$

7.2   Flow of a Non-Newtonian Fluid Over an Inclined Plate

A non-Newtonian fluid flows down an inclined plate of width W that makes an angle α with the horizontal as shown in Figure 7.2. The film extends along the plate from $z = 0$ to $z = L$.

Using our experience from Section 4.3, we postulate that $v_x = v_y = 0$, $v_z = v_z(x)$, and $P = P(x, y, z)$. Note that the equation of continuity, Eq. (A) in Table 1.1, is identically satisfied. From Table 1.2, the only non-zero shear stress component is

$$\tau_{xz} = \tau_{zx} = -\eta\frac{dv_z}{dx}. \qquad (7.2\text{-}1)$$

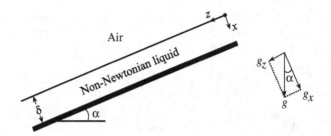

Figure 7.2   Flow of a non-Newtonian fluid over an inclined plate.

Using Eqs. (A), (B), and (C) in Table 1.5, the components of the equation of motion become

$$x\text{-component} \quad 0 = -\frac{\partial P}{\partial x} + \rho g \cos \alpha, \tag{7.2-2}$$

$$y\text{-component} \quad 0 = \frac{\partial P}{\partial y}, \tag{7.2-3}$$

$$z\text{-component} \quad 0 = -\frac{\partial P}{\partial z} - \frac{d\tau_{xz}}{dx} + \rho g \sin \alpha. \tag{7.2-4}$$

The z-component of the equation of motion for air is

$$0 = -\frac{\partial P_{\text{air}}}{\partial z} + \rho_{\text{air}} g \sin \alpha. \tag{7.2-5}$$

At $x = 0$, $P = P_{\text{air}}$ for all z. Hence,

$$\frac{\partial P}{\partial z} = \frac{\partial P_{\text{air}}}{\partial z} = \rho_{\text{air}} g \sin \alpha. \tag{7.2-6}$$

Substitution of Eq. (7.2-6) into Eq. (7.2-4) gives

$$\frac{d\tau_{xz}}{dx} = \underbrace{(\rho - \rho_{\text{air}})}_{\sim \rho} g \sin \alpha. \tag{7.2-7}$$

Integration of Eq. (7.2-7) yields

$$\tau_{xz} = \rho g x \sin \alpha + C_1. \tag{7.2-8}$$

The use of the boundary condition

$$\text{at } x = 0 \qquad \tau_{xz} = 0, \tag{7.2-9}$$

results in the following shear stress distribution:

$$\tau_{xz} = \rho g x \sin \alpha, \tag{7.2-10}$$

which is valid for both Newtonian and non-Newtonian fluids.

7.2.1 *Power-law fluid*

For a power-law fluid, the non-Newtonian viscosity is given by Eq. (7.1-6). From Eq. (A) in Table 7.1,

$$\frac{1}{2}(\dot{\gamma} : \dot{\gamma}) = \left(\frac{dv_z}{dx}\right)^2 \quad \Rightarrow \quad \dot{\gamma} = \left|\sqrt{\frac{1}{2}(\dot{\gamma} : \dot{\gamma})}\right| = \left|\frac{dv_z}{dx}\right|. \tag{7.2-11}$$

Substitution of Eq. (7.2-11) into Eq. (7.1-6) yields

$$\eta = m \left|\frac{dv_z}{dx}\right|^{n-1}. \tag{7.2-12}$$

To eliminate the absolute value sign,[4] the sign of the velocity gradient should be checked. Since v_z decreases with increasing x, $dv_z/dx < 0$ and Eq. (7.2-12) becomes

$$\eta = m \left(-\frac{dv_z}{dx}\right)^{n-1}. \tag{7.2-13}$$

Combination of Eqs. (7.2-1) and (7.2-13) gives

$$\tau_{xz} = -m \left(-\frac{dv_z}{dx}\right)^{n-1} \frac{dv_z}{dx} = m \left(-\frac{dv_z}{dx}\right)^{n}. \tag{7.2-14}$$

Substitution of Eq. (7.2-14) into Eq. (7.2-10) and rearrangement result in

$$\frac{dv_z}{dx} = -\left(\frac{\rho g \sin \alpha}{m}\right)^{s} x^{s}, \tag{7.2-15}$$

where

$$s = \frac{1}{n}. \tag{7.2-16}$$

Integration of Eq. (7.2-15) yields

$$v_z = -\left(\frac{\rho g \sin \alpha}{m}\right)^{s} \frac{x^{s+1}}{s+1} + C_2. \tag{7.2-17}$$

The use of the boundary condition

$$\text{at } x = \delta \quad v_z = 0, \tag{7.2-18}$$

---

[4]An absolute value of x is defined as

$$|x| = \begin{cases} x & \text{if } x > 0, \\ -x & \text{if } x < 0. \end{cases}$$

gives the velocity distribution as

$$v_z = \frac{\delta^{s+1}}{s+1} \left(\frac{\rho g \sin \alpha}{m} \right)^s \left[1 - \left(\frac{x}{\delta} \right)^{s+1} \right].$$

(7.2-19)

The volumetric flow rate can be determined by integrating the velocity distribution over the cross-sectional area, i.e.,

$$Q = \int_0^W \int_0^\delta v_z \, dx dy = W \int_0^\delta v_z \, dx.$$

(7.2-20)

Before substituting Eq. (7.2-19) into Eq. (7.2-20), introduction of the dimensionless distance ξ, defined by

$$\xi = \frac{x}{\delta},$$

(7.2-21)

simplifies the resulting expression as

$$Q = \frac{W \delta^{s+2}}{s+1} \left(\frac{\rho g \sin \alpha}{m} \right)^s \int_0^1 (1 - \xi^{s+1}) \, d\xi.$$

(7.2-22)

Integration leads to

$$Q = \frac{W \delta^{s+2}}{s+2} \left(\frac{\rho g \sin \alpha}{m} \right)^s.$$

(7.2-23)

For a Newtonian fluid, i.e., $m = \mu$ and $n = 1$, Eq. (7.2-23) becomes

$$Q = \frac{\rho g \delta^3 W \sin \alpha}{3\mu}.$$

(7.2-24)

When $\alpha = 90°$, Eq. (7.2-24) is identical with Eq. (4.3-16).

7.2.2   Bingham fluid

For a Bingham fluid, the non-Newtonian viscosity is given by Eq. (7.1-7). To determine the magnitude of the shear stress tensor, defined by Eq. (7.1-5), first note that

$$\boldsymbol{\tau} : \boldsymbol{\tau} = \tau_{ij} \mathbf{e}_i \mathbf{e}_j : \tau_{mn} \mathbf{e}_m \mathbf{e}_n = \tau_{ij} \tau_{mn} \underbrace{(\mathbf{e}_j \cdot \mathbf{e}_m)}_{\delta_{jm}} \underbrace{(\mathbf{e}_i \cdot \mathbf{e}_n)}_{\delta_{in}} = \tau_{ij} \tau_{ji}. \quad (7.2\text{-}25)$$

Therefore,

$$\frac{1}{2}(\boldsymbol{\tau} : \boldsymbol{\tau}) = \frac{1}{2}(\tau_{xz}\tau_{zx} + \tau_{zx}\tau_{xz}) = \tau_{xz}^2.$$

(7.2-26)

Since τ_{xz} is positive, then

$$\tau = |\tau_{xz}| = \tau_{xz}. \tag{7.2-27}$$

The magnitude of the shear rate is determined from Eq. (A) in Table 7.1 as

$$\dot{\gamma} = \left| \frac{dv_z}{dx} \right|. \tag{7.2-28}$$

Since v_z decreases with increasing x, then $dv_z/dx < 0$ and Eq. (7.2-28) becomes

$$\dot{\gamma} = -\frac{dv_z}{dx}. \tag{7.2-29}$$

Equation (7.2-10) indicates that $\tau_{xz} = 0$ at $x = 0$. Let x_o be the distance at which $\tau = \tau_o$. The value of x_o is determined from Eq. (7.2-10) as

$$x_o = \frac{\tau_o}{\rho g \sin \alpha}. \tag{7.2-30}$$

Therefore, the flow region is divided into two zones, namely, region I ($x_o \le x \le \delta$), in which $\tau \ge \tau_o$, and region II ($0 \le x \le x_o$), in which $\tau \le \tau_o$.

- **Region I** ($\tau \ge \tau_o$)
  From Eq. (7.1-7b),

$$\tau = \mu_o + \frac{\tau_o}{\dot{\gamma}}. \tag{7.2-31}$$

Substitution of Eqs. (7.2-27) and (7.2-29) into Eq. (7.2-31) gives

$$\tau_{xz} = -\left(\mu_o + \frac{\tau_o}{-dv_z^I/dx} \right) \frac{dv_z^I}{dx} = -\mu_o \frac{dv_z^I}{dx} + \tau_o. \tag{7.2-32}$$

The use of Eq. (7.2-32) in Eq. (7.2-10) and rearrangement yield

$$\frac{dv_z^I}{dx} = -\frac{\rho g \sin \alpha}{\mu_o} x + \frac{\tau_o}{\mu_o}. \tag{7.2-33}$$

Integration of Eq. (7.2-33) leads to

$$v_z^I = -\frac{\rho g \sin \alpha}{2\mu_o} x^2 + \frac{\tau_o}{\mu_o} x + C_3. \tag{7.2-34}$$

The use of the boundary condition

$$\text{at } x = \delta \qquad v_z = 0, \tag{7.2-35}$$

gives the velocity distribution as

$$v_z^I = \frac{\rho g \delta^2 \sin \alpha}{2\mu_o} \left[1 - \left(\frac{x}{\delta} \right)^2 \right] - \frac{\tau_o \delta}{\mu_o} \left(1 - \frac{x}{\delta} \right) \qquad x_o \le x \le \delta. \tag{7.2-36}$$

- **Region II** $(\tau \leq \tau_o)$

Since $\eta = \infty$ from Eq. (7.1-7a), rearrangement of Eq. (7.2-1) in the form

$$\frac{dv_z^{II}}{dx} = -\frac{1}{\eta} \tau_{xz}, \qquad (7.2\text{-}37)$$

implies that

$$\frac{dv_z^{II}}{dx} = 0. \qquad (7.2\text{-}38)$$

In other words, region II is a plug flow region in which the velocity is uniform. Evaluation of Eq. (7.2-36) at x_o gives the velocity as

$$v_z^{II} = \frac{\rho g \delta^2 \sin \alpha}{2\mu_o} \left[1 - \left(\frac{x_o}{\delta}\right)^2 \right] - \frac{\tau_o \delta}{\mu_o} \left(1 - \frac{x_o}{\delta} \right) \qquad 0 \leq x \leq x_o.$$

$$(7.2\text{-}39)$$

A representative velocity distribution is shown in Figure 7.3.

The volumetric flow rate is given by

$$Q = \int_0^W \int_0^\delta v_z \, dx \, dy = W \int_0^\delta v_z \, dx. \qquad (7.2\text{-}40)$$

By choosing

$$u = v_z \qquad \text{and} \qquad dv = dx, \qquad (7.2\text{-}41)$$

integration by parts[5] gives

$$\frac{Q}{W} = \underbrace{(xv_z)\Big|_0^\delta}_{0} - \int_0^\delta x \left(\frac{dv_z}{dx}\right) dx. \qquad (7.2\text{-}42)$$

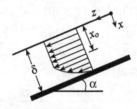

Figure 7.3   A representative velocity distribution for the flow of a Bingham fluid over an inclined plate.

---

[5]The formula for the integration by parts is given as

$$\int u \, dv = uv - \int v \, du.$$

Since $dv_z/dx = 0$ when $0 \leq x \leq x_o$, Eq. (7.2-42) simplifies to

$$\frac{Q}{W} = -\int_{x_o}^{\delta} x \left(\frac{dv_z^I}{dx} \right) dx. \tag{7.2-43}$$

The use of Eq. (7.2-35) in Eq. (7.2-43) leads to

$$\frac{Q}{W} = \frac{\delta^2 \tau_\delta}{3\mu_o} \left[1 - \frac{3}{2} \left(\frac{\tau_o}{\tau_\delta} \right) + \frac{1}{2} \left(\frac{\tau_o}{\tau_\delta} \right)^3 \right], \tag{7.2-44}$$

where

$$\tau_\delta = \rho g \delta \sin \alpha. \tag{7.2-45}$$

For a Newtonian fluid, i.e., $\mu_o = \mu$ and $\tau_o = 0$, Eq. (7.2-44) reduces to Eq. (7.2-24).

7.3   Flow of an Ellis Fluid in a Circular Tube

Consider a steady, laminar flow of an incompressible Ellis fluid through an inclined tube of radius R as shown in Figure 5.3. Using our experience from Section 5.2, we postulate that $v_r = v_\theta = 0$, $v_z = v_z(r)$, and $P = P(r, \theta, z)$. Note that the equation of continuity, Eq. (B) in Table 1.1, is identically satisfied. From Table 1.3, the only non-zero shear stress component is

$$\tau_{rz} = \tau_{zr} = -\eta \frac{dv_z}{dr}. \tag{7.3-1}$$

Using Eqs. (A), (B), and (C) in Table 1.6, the components of the equation of motion become

$$r\text{-component} \quad 0 = -\frac{\partial P}{\partial r} + \rho g \cos \alpha \sin \theta, \tag{7.3-2}$$

$$\theta\text{-component} \quad 0 = -\frac{1}{r} \frac{\partial P}{\partial \theta} + \rho g \cos \alpha \cos \theta, \tag{7.3-3}$$

$$z\text{-component} \quad 0 = -\frac{\partial P}{\partial z} - \frac{1}{r} \frac{d}{dr} (r \tau_{rz}) - \rho g \sin \alpha. \tag{7.3-4}$$

Defining the modified pressure as

$$\mathcal{P} = P + \rho g z \sin \alpha. \tag{7.3-5}$$

reduces Eqs. (7.3-2)–(7.3-4) to the following form:

$$\frac{\partial P}{\partial r} = \rho g \cos \alpha \sin \theta, \tag{7.3-6}$$

$$\frac{\partial P}{\partial \theta} = \rho g r \cos \alpha \cos \theta, \tag{7.3-7}$$

$$\frac{\partial P}{\partial z} = -\frac{1}{r}\frac{d}{dr}\left(r\tau_{rz}\right). \tag{7.3-8}$$

The analysis presented in Section 5.2 indicates that the pressure distribution is given by Eq. (5.2-25), i.e.,

$$P - P_o = -\left(\frac{P_o - P_L}{L}\right)z + \rho g \cos \alpha (r \sin \theta + R). \tag{7.3-9}$$

The use of Eq. (7.3-9) in Eq. (7.3-8) results in

$$\frac{1}{r}\frac{d}{dr}(r\tau_{rz}) = \frac{P_o - P_L}{L}. \tag{7.3-10}$$

Equation (7.3-10) holds for both Newtonian and non-Newtonian fluids. Integration of Eq. (7.3-10) gives

$$\tau_{rz} = \left(\frac{P_o - P_L}{2L}\right)r + \frac{C_1}{r}. \tag{7.3-11}$$

Since τ_{rz} is finite at the tube center, $C_1 = 0$ and Eq. (7.3-11) simplifies to

$$\tau_{rz} = \left(\frac{P_o - P_L}{2L}\right)r. \tag{7.3-12}$$

The value of the shear stress at the wall, τ_R, is

$$\tau_R = \left(\frac{P_o - P_L}{2L}\right)R. \tag{7.3-13}$$

Using Eqs. (7.3-12) and (7.3-13), we obtain

$$\tau_{rz} = \tau_R \left(\frac{r}{R}\right). \tag{7.3-14}$$

For an Ellis fluid, the non-Newtonian viscosity is given by Eq. (7.1-8). The use of Eq. (7.2-25) gives

$$\frac{1}{2}(\boldsymbol{\tau} : \boldsymbol{\tau}) = \frac{1}{2}(\tau_{rz}^2 + \tau_{zr}^2) = \tau_{rz}^2. \tag{7.3-15}$$

Since τ_{rz} is positive,

$$\tau = |\tau_{rz}| = \tau_{rz}. \tag{7.3-16}$$

and Eq. (7.1-8) becomes

$$\frac{1}{\eta} = \frac{1}{\eta_o} + \frac{1}{\eta_o}\left(\frac{\tau_{rz}}{\tau_{1/2}}\right)^{\alpha-1}. \tag{7.3-17}$$

Substitution of Eq. (7.3-17) into Eq. (7.3-1) yields

$$\frac{dv_z}{dr} = -\left[\frac{1}{\eta_o} + \frac{1}{\eta_o}\left(\frac{\tau_{rz}}{\tau_{1/2}}\right)^{\alpha-1}\right]\tau_{rz}. \tag{7.3-18}$$

With the help of Eq. (7.3-14), Eq. (7.3-18) takes the following form:

$$\frac{dv_z}{dr} = -\frac{\tau_R}{R\eta_o}\left[r + \left(\frac{\tau_R}{R\tau_{1/2}}\right)^{\alpha-1} r^\alpha\right]. \tag{7.3-19}$$

Integration of Eq. (7.3-19) gives

$$v_z = -\frac{\tau_R}{R\eta_o}\left[\frac{r^2}{2} + \left(\frac{\tau_R}{R\tau_{1/2}}\right)^{\alpha-1}\frac{r^{\alpha+1}}{\alpha+1}\right] + C_2. \tag{7.3-20}$$

The use of the boundary condition

$$\text{at } r = R \qquad v_z = 0, \tag{7.3-21}$$

gives the velocity distribution as

$$v_z = \frac{\tau_R R}{2\eta_o}\left\{1 - \left(\frac{r}{R}\right)^2 + \frac{2}{\alpha+1}\left(\frac{\tau_R}{\tau_{1/2}}\right)^{\alpha-1}\left[1 - \left(\frac{r}{R}\right)^{\alpha+1}\right]\right\}. \tag{7.3-22}$$

The volumetric flow rate is

$$Q = \int_0^{2\pi}\int_0^R v_z r\, dr\, d\theta = 2\pi\int_0^R v_z r\, dr. \tag{7.3-23}$$

In terms of the dimensionless distance ξ, defined by

$$\xi = \frac{r}{R}, \tag{7.3-24}$$

Eq. (7.3-23) becomes

$$Q = 2\pi R^2 \int_0^1 v_z \xi\, d\xi. \tag{7.3-25}$$

Substitution of Eq. (7.3-22) into Eq. (7.3-25) and integration lead to

$$Q = \frac{\pi R^3 \tau_R}{4\eta_o} \left[1 + \frac{4}{\alpha+3} \left(\frac{\tau_R}{\tau_{1/2}} \right)^{\alpha-1} \right]. \qquad (7.3\text{-}26)$$

For a Newtonian fluid, i.e., $\eta_o = \mu$ and $\alpha = 1$, Eq. (7.3-26) reduces to Eq. (5.2-35).

Numerical Example   For an aqueous solution of carboxymethylcellulose (CMC), Ashare *et al.* (1965) reported the following Ellis model parameters:

$$\eta_o = 2.27 \ \text{Pa} \cdot \text{s} \qquad \tau_{1/2} = 152 \ \text{Pa} \qquad \alpha = 3.$$

It is required to transport this CMC solution at a volumetric flow rate of $280 \ \text{cm}^3/\text{s}$ through a tube of radius 0.5 cm. Calculate the pressure gradient to accomplish this task.

Solution   Substitution of the numerical values into Eq. (7.3-26) gives

$$280 \times 10^{-6} = \frac{\pi (0.5 \times 10^{-2})^3 \tau_R}{(4)(2.27)} \left[1 + \frac{4}{6} \left(\frac{\tau_R}{152} \right)^2 \right] \quad \Rightarrow \quad \tau_R = 588.6 \ \text{Pa}.$$

Thus, the pressure gradient is calculated from Eq. (7.3-13) as

$$\frac{\mathcal{P}_o - \mathcal{P}_L}{L} = \frac{2\tau_R}{R} = \frac{(2)(588.6)}{0.5 \times 10^{-2}} = 2.354 \times 10^5 \ \text{Pa/m} = 2.32 \ \text{atm/m}.$$

7.4   Blood Flow in a Tube

Blood is a suspension of red cells (particles about 8 μm in diameter) in an aqueous solution called plasma. Plasma behaves as a Newtonian fluid, whereas the combination of red cells and plasma turns out to be described adequately by the power-law fluid.

It has also been documented that as blood flows under steady conditions in a long and thin circular tube of radius R, the red cells tend to migrate away from the wall, leaving a red cell-free layer next to the tube wall with a thickness of δ. Therefore, both Newtonian and non-Newtonian flow areas are present at a given tube cross-section.

For flow through a circular tube, Eq. (7.3-11) gives the shear stress distribution for Newtonian and non-Newtonian fluids in the form

$$\tau_{rz} = \left(\frac{\mathcal{P}_o - \mathcal{P}_L}{2L} \right) r + \frac{C_1}{r}. \qquad (7.4\text{-}1)$$

In the power-law region, i.e., $0 \leq r \leq R-\delta$, the shear stress distribution, Eq. (7.4-1), is expressed as

$$\tau_{rz}^{B} = \left(\frac{\mathcal{P}_o - \mathcal{P}_L}{2L}\right) r + \frac{C_1^{B}}{r}, \qquad (7.4-2)$$

in which the superscript B stands for blood. Since the shear stress must have a finite value at the tube center, i.e., $r = 0$, then $C_1^{B} = 0$ and Eq. (7.4-2) simplifies to

$$\tau_{rz}^{B} = \left(\frac{\mathcal{P}_o - \mathcal{P}_L}{2L}\right) r. \qquad (7.4-3)$$

For a power-law fluid, the non-Newtonian viscosity is given by Eq. (7.1-6). From Eq. (B) in Table 7.1,

$$\frac{1}{2}\left(\dot{\gamma} : \dot{\gamma}\right) = \left(\frac{dv_z^{B}}{dr}\right)^2 \quad \Rightarrow \quad \dot{\gamma} = \left|\sqrt{\frac{1}{2}\left(\dot{\gamma} : \dot{\gamma}\right)}\right| = \left|\frac{dv_z^{B}}{dr}\right|. \qquad (7.4-4)$$

Substitution of Eq. (7.4-4) into Eq. (7.1-6) yields

$$\eta = m \left|\frac{dv_z^{B}}{dr}\right|^{n-1}. \qquad (7.4-5)$$

Therefore, the shear stress expression takes the following form:

$$\tau_{rz}^{B} = -\eta \frac{dv_z^{B}}{dr} = -m \left|\frac{dv_z^{B}}{dr}\right|^{n-1} \frac{dv_z^{B}}{dr}. \qquad (7.4-6)$$

In the power-law region, velocity decreases with increasing r value. In other words, the velocity gradient is negative. Therefore, Eq. (7.4-6) takes the form

$$\tau_{rz}^{B} = m \left(-\frac{dv_z^{B}}{dr}\right)^{n}. \qquad (7.4-7)$$

Substitution of Eq. (7.4-7) into Eq. (7.4-3) and rearrangement give

$$\frac{dv_z^{B}}{dr} = -\left(\frac{\mathcal{P}_o - \mathcal{P}_L}{2mL}\right)^{s} r^{s}, \qquad (7.4-8)$$

where

$$s = \frac{1}{n}. \qquad (7.4-9)$$

Integration of Eq. (7.4-8) leads to

$$v_z^B = -\frac{1}{s+1}\left(\frac{\mathcal{P}_o - \mathcal{P}_L}{2mL}\right)^s r^{s+1} + C_2. \tag{7.4-10}$$

In the Newtonian region, i.e., $R - \delta \leq r \leq R$, the shear stress distribution, Eq. (7.4-1), is expressed as

$$\tau_{rz}^P = \left(\frac{\mathcal{P}_o - \mathcal{P}_L}{2L}\right)r + \frac{C_1^P}{r}, \tag{7.4-11}$$

in which the superscript P stands for plasma. The shear stress is a continuous function at the blood–plasma interface, i.e.,

$$\text{at } r = R - \delta \qquad \tau_{rz}^B = \tau_{rz}^P, \tag{7.4-12}$$

Substitution of Eqs. (7.4-3) and (7.4-11) into Eq. (7.4-12) gives $C_1^P = 0$ and Eq. (7.4-11) reduces to

$$\tau_{rz}^P = \left(\frac{\mathcal{P}_o - \mathcal{P}_L}{2L}\right)r. \tag{7.4-13}$$

For a Newtonian fluid,

$$\tau_{rz}^P = -\mu^P \frac{dv_z^P}{dr}. \tag{7.4-14}$$

Substitution of Eq. (7.4-14) into Eq. (7.4-13) and integration give

$$v_z^P = -\left(\frac{\mathcal{P}_o - \mathcal{P}_L}{4\mu^P L}\right)r^2 + C_3. \tag{7.4-15}$$

The use of the boundary condition

$$\text{at } r = R \qquad v_z^P = 0, \tag{7.4-16}$$

gives the velocity distribution in the plasma region as

$$v_z^P = \left(\frac{\mathcal{P}_o - \mathcal{P}_L}{4\mu^P L}\right)R^2\left[1 - \left(\frac{r}{R}\right)^2\right]. \tag{7.4-17}$$

Velocities are equal to each other at the blood–plasma interface, i.e.,

$$\text{at } r = R - \delta \qquad v_z^B = v_z^P. \tag{7.4-18}$$

Substitution of Eqs. (7.4-10) and (7.4-17) into Eq. (7.4-18) yields the velocity distribution in the blood region as

$$v_z^B = \frac{1}{s+1} \left(\frac{\mathcal{P}_o - \mathcal{P}_L}{2mL} \right)^s (R - \delta)^{s+1} \left[1 - \left(\frac{r}{R - \delta} \right)^{s+1} \right]$$

$$+ \frac{(\mathcal{P}_o - \mathcal{P}_L)R^2}{4\mu^P L} \left[2 \left(\frac{\delta}{R} \right) - \left(\frac{\delta}{R} \right)^2 \right]. \tag{7.4-19}$$

The volumetric flow rate is given as

$$Q = \int_0^{2\pi} \int_0^{R-\delta} v_z^B r \, dr \, d\theta + \int_0^{2\pi} \int_{R-\delta}^R v_z^P r \, dr \, d\theta$$

$$= 2\pi \left(\int_0^{R-\delta} v_z^B r \, dr + \int_{R-\delta}^R v_z^P r \, dr \right). \tag{7.4-20}$$

Substitution of Eqs. (7.4-17) and (7.4-19) into Eq. (7.4-20) and integration give the volumetric flow rate as

$$Q = \frac{\pi R^{s+3}}{s+3} \left(\frac{\mathcal{P}_o - \mathcal{P}_L}{2mL} \right)^s \left(\frac{R - \delta}{R} \right)^{s+3}$$

$$+ \frac{\pi (\mathcal{P}_o - \mathcal{P}_L)R^4}{8\mu^P L} \left[1 - \left(\frac{R - \delta}{R} \right)^4 \right]. \tag{7.4-21}$$

7.4.1 *Investigation of limiting cases*

- **Newtonian fluid**

When

$$n = 1 \qquad m = \mu^P = \mu \qquad \delta = R, \tag{7.4-22}$$

Eq. (7.4-21) reduces to

$$Q = \frac{\pi (\mathcal{P}_o - \mathcal{P}_L)R^4}{8\mu L}, \tag{7.4-23}$$

which is identical with Eq. (5.2-35).

- **Power-law fluid**

When $\delta = 0$, the velocity distribution, Eq. (7.4-19), simplifies to

$$v_z = \frac{1}{s+1} \left(\frac{\mathcal{P}_o - \mathcal{P}_L}{2mL} \right)^s R^{s+1} \left[1 - \left(\frac{r}{R} \right)^{s+1} \right]. \tag{7.4-24}$$

On the other hand, the volumetric flow rate expression, Eq. (7.4-21), becomes

$$Q = \frac{\pi R^{s+3}}{s+3} \left(\frac{\mathcal{P}_o - \mathcal{P}_L}{2mL} \right)^s.$$

(7.4-25)

7.5   Flow of a Power-Law Fluid in an Annulus

Consider steady, laminar flow of an incompressible power-law fluid through an inclined concentric annulus as shown in Figure 5.6. Following the procedure given in Section 7.3, the shear stress distribution is given by

$$\frac{1}{r} \frac{d}{dr} (r\tau_{rz}) = \frac{\mathcal{P}_o - \mathcal{P}_L}{L},$$

(7.5-1)

where

$$\tau_{rz} = -\eta \frac{dv_z}{dr}.$$

(7.5-2)

Integration of Eq. (7.5-1) gives

$$\tau_{rz} = \left(\frac{\mathcal{P}_o - \mathcal{P}_L}{2L} \right) r + \frac{C_1}{r}.$$

(7.5-3)

For the problem at hand, $\kappa R \leq r \leq R$. Since $r = 0$ is outside of the flow domain, the constant C_1 cannot be set equal to zero. If the velocity reaches its maximum value at $r = \beta R$, then Eq. (7.5-2) gives the boundary condition as[6]

$$\text{at } r = \beta R \qquad \tau_{rz} = 0.$$

(7.5-4)

Thus, the shear stress distribution takes the form

$$\tau_{rz} = \frac{(\mathcal{P}_o - \mathcal{P}_L)R}{2L} \left(\xi - \frac{\beta^2}{\xi} \right),$$

(7.5-5)

where the dimensionless distance ξ is defined by

$$\xi = \frac{r}{R}.$$

(7.5-6)

---

[6]It should be kept in mind that β is an unknown that will be determined at a later stage.

Equation (7.5-5) is valid for both Newtonian and non-Newtonian fluids.

From Eq. (B) in Table 7.1,

$$\frac{1}{2}(\dot{\gamma}:\dot{\gamma}) = \left(\frac{dv_z}{dr}\right)^2 \quad \Rightarrow \quad \dot{\gamma} = \left|\sqrt{\frac{1}{2}(\dot{\gamma}:\dot{\gamma})}\right| = \left|\frac{dv_z}{dr}\right|. \tag{7.5-7}$$

Substitution of Eq. (7.5-7) into Eq. (7.1-6) yields

$$\eta = m\left|\frac{dv_z}{dr}\right|^{n-1}. \tag{7.5-8}$$

The use of Eq. (7.5-8) in Eq. (7.5-2) leads to the following expression for the shear stress:

$$\tau_{rz} = -m\left|\frac{dv_z}{dr}\right|^{n-1}\frac{dv_z}{dr}. \tag{7.5-9}$$

In the region $\kappa \leq \xi \leq \beta$, the velocity gradient is positive, i.e., $dv_z/dr > 0$. Therefore, Eq. (7.5-9) becomes

$$\tau_{rz} = -m\left(\frac{dv_z}{dr}\right)^{n-1}\left(\frac{dv_z}{dr}\right) = -m\left(\frac{dv_z}{dr}\right)^n$$

$$= -m\left(\frac{1}{R}\frac{dv_z}{d\xi}\right)^n \quad \kappa \leq \xi \leq \beta. \tag{7.5-10}$$

In the region $\beta \leq \xi \leq 1$, the velocity gradient is negative, i.e., $dv_z/dr < 0$. In this case, Eq. (7.5-9) becomes

$$\tau_{rz} = -m\left(-\frac{dv_z}{dr}\right)^{n-1}\left(\frac{dv_z}{dr}\right) = m\left(-\frac{dv_z}{dr}\right)^n$$

$$= m\left(-\frac{1}{R}\frac{dv_z}{d\xi}\right)^n \quad \beta \leq \xi \leq 1. \tag{7.5-11}$$

- **Velocity distribution for $\kappa \leq \xi \leq \beta$**

Substitution of Eq. (7.5-10) into Eq. (7.5-5) and integration yield

$$v_z = R\left[\frac{(\mathcal{P}_o - \mathcal{P}_L)R}{2mL}\right]^s \int_\kappa^\xi u^{-s}\left(\beta^2 - u^2\right)^s du, \tag{7.5-12}$$

where

$$s = \frac{1}{n}. \tag{7.5-13}$$

The term u in Eq. (7.5-12) is a dummy variable of integration.

- **Velocity distribution for** $\beta \leq \xi \leq 1$

  Substitution of Eq. (7.5-11) into Eq. (7.5-5) and integration yield

  $$v_z = R \left[\frac{(\mathcal{P}_o - \mathcal{P}_L)R}{2\,mL} \right]^s \int_\xi^1 u^{-s} \left(u^2 - \beta^2 \right)^s du. \qquad (7.5\text{-}14)$$

- **Determination of** β

  The parameter β in Eqs. (7.5-12) and (7.5-14) is an unknown representing the location of maximum velocity. Equating these equations at $\xi = \beta$ leads to

  $$\int_\kappa^\beta u^{-s} \left(\beta^2 - u^2 \right)^s du = \int_\beta^1 u^{-s} \left(u^2 - \beta^2 \right)^s du. \qquad (7.5\text{-}15)$$

  The solution of this equation, which is obviously not straightforward, gives β as a function of κ (related to flow geometry) and s (related to fluid type).

7.5.1   *Volumetric flow rate*

The volumetric flow rate is given by

$$Q = \int_0^{2\pi} \int_{\kappa R}^R v_z r \, dr d\theta = 2\pi R^2 \int_\kappa^1 v_z \xi \, d\xi$$

$$= 2\pi R^2 \left(\int_\kappa^\beta v_z \xi \, d\xi + \int_\beta^1 v_z \xi \, d\xi \right). \qquad (7.5\text{-}16)$$

Substitution of Eqs. (7.5-12) and (7.5-14) into Eq. (7.5-16) yields

$$Q = 2\pi R^3 \left[\frac{(\mathcal{P}_o - \mathcal{P}_L)R}{2mL} \right]^s$$

$$\times \left[\underbrace{\int_\kappa^\beta \int_\kappa^\xi u^{-s} \left(\beta^2 - u^2 \right)^s \xi \, du \, d\xi}_{A} + \underbrace{\int_\beta^1 \int_\xi^1 u^{-s} \left(u^2 - \beta^2 \right)^s \xi \, du \, d\xi}_{B} \right]. \qquad (7.5\text{-}17)$$

Figure 7.4 shows how to change the order of integrations in Eq. (7.5-17). Thus, Eq. (7.5-17) takes the following form:

$$Q = 2\pi R^3 \left[\frac{(\mathcal{P}_o - \mathcal{P}_L)R}{2mL} \right]^s$$

$$\times \left[\int_\kappa^\beta \int_u^\beta u^{-s}(\beta^2 - u^2)^s \xi \, d\xi \, du + \int_\beta^1 \int_\beta^u u^{-s}(u^2 - \beta^2)^s \xi \, d\xi \, du \right].$$

(7.5-18)

Carrying out the integrations in Eq. (7.5-18) with respect to ξ gives

$$Q = \pi R^3 \left[\frac{(\mathcal{P}_o - \mathcal{P}_L)R}{2mL} \right]^s$$

$$\times \left[\int_\kappa^\beta u^{-s}(\beta^2 - u^2)^{s+1} \, du + \int_\beta^1 u^{-s}(u^2 - \beta^2)^{s+1} \, du \right]. \quad (7.5\text{-}19)$$

Equation (7.5-19) was first obtained by Fredrickson and Bird (1958). Later, Hanks and Larsen (1979) evaluated the integrals in Eq. (7.5-19) analytically for arbitrary values of s by the following procedure.

In the first integral on the right-hand side of Eq. (7.5-19), by choosing

$$u = \int_\kappa^\xi \left(\frac{\beta^2}{u} - u \right)^s du \quad \text{and} \quad dv = \xi \, d\xi, \quad (7.5\text{-}20)$$

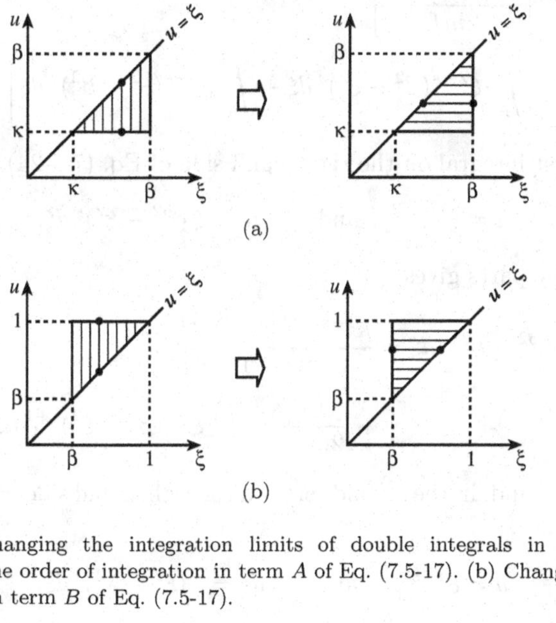

(a)

(b)

Figure 7.4 Changing the integration limits of double integrals in Eq. (7.5-17). (a) Changing the order of integration in term A of Eq. (7.5-17). (b) Changing the order of integration in term B of Eq. (7.5-17).

integration by parts gives

$$\int_\kappa^\beta \left[\int_\kappa^\xi \left(\frac{\beta^2}{u} - u \right)^s du \right] \xi \, d\xi = \frac{\beta^2}{2} \int_\kappa^\beta \left(\frac{\beta^2}{u} - u \right)^s du$$

$$- \frac{1}{2} \int_\kappa^\beta \xi^{2-s} (\beta^2 - \xi^2)^s d\xi. \qquad (7.5\text{-}21)$$

In the second integral on the right-hand side of Eq. (7.5-19), by choosing

$$u = \int_\xi^1 \left(u - \frac{\beta^2}{u} \right)^s du \qquad \text{and} \qquad dv = \xi \, d\xi, \qquad (7.5\text{-}22)$$

integration by parts gives

$$\int_\beta^1 \left[\int_\xi^1 \left(u - \frac{\beta^2}{u} \right)^s du \right] \xi \, d\xi = -\frac{\beta^2}{2} \int_\beta^1 \left(u - \frac{\beta^2}{u} \right)^s du$$

$$+ \frac{1}{2} \int_\beta^1 \xi^{2-s} (\xi^2 - \beta^2)^s d\xi. \qquad (7.5\text{-}23)$$

Substitution of Eqs. (7.5-21) and (7.5-23) into Eq. (7.5-19) and simplification of the resulting expression using Eq. (7.5-15) yield an alternate expression for the volumetric flow rate as

$$Q = \pi R^3 \left[\frac{(\mathcal{P}_o - \mathcal{P}_L)R}{2mL} \right]^s$$

$$\times \left[-\int_\kappa^\beta \xi^{2-s} (\beta^2 - \xi^2)^s \, d\xi + \int_\beta^1 \xi^{2-s} (\xi^2 - \beta^2)^s \, d\xi \right]. \qquad (7.5\text{-}24)$$

In the first integral on the right-hand side of Eq. (7.5-24), by choosing

$$u = \xi^{1-s} \qquad \text{and} \qquad dv = \xi \, (\beta^2 - \xi^2)^s \, d\xi, \qquad (7.5\text{-}25)$$

integration by parts gives

$$\int_\kappa^\beta \xi^{2-s} (\beta^2 - \xi^2)^s \, d\xi = \frac{\kappa^{1-s} (\beta^2 - \kappa^2)^{s+1}}{2(s+1)}$$

$$+ \frac{1-s}{2(s+1)} \int_\kappa^\beta \xi^{-s} (\beta^2 - \xi^2)^{s+1} \, d\xi. \qquad (7.5\text{-}26)$$

On the other hand, in the second term on the right-hand side of Eq. (7.5-24), by choosing

$$u = \xi^{1-s} \qquad \text{and} \qquad dv = \xi \, (\xi^2 - \lambda^2)^s \, d\xi, \qquad (7.5\text{-}27)$$

integration by parts gives

$$\int_\beta^1 \xi^{2-s}(\xi^2 - \beta^2)^s \, d\xi = \frac{(1-\beta^2)^{s+1}}{2(s+1)}$$

$$- \frac{1-s}{2(s+1)} \int_\beta^1 \xi^{-s}(\xi^2 - \beta^2)^{s+1} \, d\xi. \quad (7.5\text{-}28)$$

Substitution of Eqs. (7.5-26) and (7.5-28) into Eq. (7.5-24) gives

$$Q = \pi R^3 \left[\frac{(\mathcal{P}_o - \mathcal{P}_L)R}{2mL} \right]^s \left\{ \left[-\frac{\kappa^{1-s}(\beta^2 - \kappa^2)^{s+1}}{2(s+1)} + \frac{(1-\beta^2)^{s+1}}{2(s+1)} \right] \right.$$

$$\left. - \frac{1-s}{2(s+1)} \left[\int_\kappa^\beta \xi^{-s}(\beta^2 - \xi^2)^{s+1} \, d\xi + \int_\beta^1 \xi^{-s}(\xi^2 - \beta^2)^{s+1} \, d\xi \right] \right\}.$$

$$(7.5\text{-}29)$$

Clearly, Eqs. (7.5-17) and (7.5-29) must be equivalent results. Equating these two equations we get

$$\int_\kappa^\beta u^{-s}(\beta^2 - u^2)^{s+1} \, du + \int_\beta^1 u^{-s}(u^2 - \beta^2)^{s+1} \, du$$

$$= \frac{(1-\beta^2)^{s+1} - \kappa^{1-s}(\beta^2 - \kappa^2)^{s+1}}{s+3}. \quad (7.5\text{-}30)$$

so that the volumetric flow rate becomes

$$Q = \frac{\pi R^3}{s+3} \left[\frac{(\mathcal{P}_o - \mathcal{P}_L)R}{2\,mL} \right]^s \left[(1-\beta^2)^{s+1} - \kappa^{1-s}(\beta^2 - \kappa^2)^{s+1} \right]. \quad (7.5\text{-}31)$$

• **Simplification of Eq. (7.5-31) for a Newtonian fluid**

For a Newtonian fluid, $n = s = 1$ and $m = \mu$. Thus, Eq. (7.5-31) simplifies to

$$Q = \frac{\pi(\mathcal{P}_o - \mathcal{P}_L)R^4}{8\mu L} \left[1 - \kappa^4 - 2\beta^2(1 - \kappa^2) \right]. \quad (7.5\text{-}32)$$

On the other hand, Eq. (7.5-15) becomes

$$\int_\kappa^\beta \left(\frac{\beta^2}{u} - u \right) du = \int_\beta^1 \left(u - \frac{\beta^2}{u} \right) du. \quad (7.5\text{-}33)$$

Integration leads to

$$2\beta^2 = \frac{1 - \kappa^2}{\ln(1/\kappa)}. \quad (7.5\text{-}34)$$

Substitution of Eq. (7.5-34) into Eq. (7.5-32) gives

$$Q = \frac{\pi (\mathcal{P}_o - \mathcal{P}_L) R^4}{8\mu L} \left[1 - \kappa^4 - \frac{(1 - \kappa^2)^2}{\ln(1/\kappa)} \right], \qquad (7.5\text{-}35)$$

which is identical with Eq. (5.3-14).

Numerical Example A polymer solution flows through a concentric annulus at a volumetric flow rate of $0.2 \text{ m}^3/\text{min}$. The polymer solution has the following power-law model parameters:

$$n = 0.4 \qquad \text{and} \qquad m = 3.1 \text{ Pa} \cdot \text{s}^{0.4}.$$

Calculate the applied pressure gradient if the diameters of the inner and outer pipes are 1.1 cm and 2 cm, respectively.

Solution The values of s and κ are

$$s = \frac{1}{0.4} = 2.5 \qquad \text{and} \qquad \kappa = \frac{1.1}{2} = 0.55.$$

Substitution of the numerical values into Eq. (7.5-15) gives

$$\int_{0.55}^{\beta} u^{-2.5} \left(\beta^2 - u^2 \right)^{2.5} du = \int_{\beta}^{1} u^{-2.5} \left(u^2 - \beta^2 \right)^{2.5} du.$$

Numerical solution of the above equation by either MATLAB or Mathcad gives

$$\beta = 0.756.$$

Rearrangement of Eq. (7.5-31) results in

$$\frac{\mathcal{P}_o - \mathcal{P}_L}{L} = 2m \left[\frac{(s+3)Q}{\pi R^{s+3} \chi} \right]^{1/s},$$

where

$$\chi = (1 - \beta^2)^{s+1} - \kappa^{1-s} (\beta^2 - \kappa^2)^{s+1}.$$

Substitution of the numerical values leads to

$$\frac{\mathcal{P}_o - \mathcal{P}_L}{L} = (2)(3.1) \left[\frac{(2.5 + 3)(0.2/60)}{\pi (0.01)^{2.5+1} (0.027)} \right]^{0.4} = 8.439 \times 10^4 \text{ Pa/m}$$

$$= 0.83 \text{ atm/m}.$$

7.6 Couette Flow of a Bingham Fluid Between Concentric Cylinders

A Bingham fluid is flowing tangentially in the region between two concentric cylinders having inner and outer radii of κR and R, respectively. The cylinder at $r = \kappa R$ is fixed and that at $r = R$ moves with an angular velocity Ω because of the applied torque T_z.

Using our experience from Section 4.2, we postulate that $v_r = v_z = 0$, $v_\theta = v_\theta(r)$, and $P = P(r, z)$. From Table 1.3, the only non-zero shear stress component is

$$\tau_{r\theta} = \tau_{\theta r} = -\eta r \frac{d}{dr}\left(\frac{v_\theta}{r}\right). \tag{7.6-1}$$

From Eq. (B) in Table 1.6, the θ-component of the equation of motion simplifies to

$$0 = \frac{1}{r^2}\frac{d}{dr}(r^2\tau_{r\theta}) \quad \Rightarrow \quad \tau_{r\theta} = \frac{C_1}{r^2}, \tag{7.6-2}$$

which is valid for both Newtonian and non-Newtonian fluids.

Now, let us express the integration constant C_1 in terms of the applied torque T_z. The z-component of the torque is given by Eq. (4.2-29), i.e.,

$$T_z = \int_A \mathbf{e}_z \cdot [\mathbf{r} \times (\mathbf{n} \cdot \boldsymbol{\pi})]\, dA. \tag{7.6-3}$$

To calculate the torque exerted by the outer cylinder on the fluid, the unit normal vector should be directed from the outer cylinder to the fluid, i.e., $\mathbf{n} = -\mathbf{e}_r$. From Eq. (4.2-28), the position vector in the cylindrical coordinate system is $\mathbf{r} = r\mathbf{e}_r + z\mathbf{e}_z$. Thus, the integrand in Eq. (7.6-3) is

$$\begin{aligned}
\mathbf{e}_z \cdot [\mathbf{r} \times (\mathbf{n} \cdot \boldsymbol{\pi})] &= \mathbf{e}_z \cdot [(r\mathbf{e}_r + z\mathbf{e}_z) \times (-P\mathbf{e}_r - \mathbf{e}_r \cdot \tau_{ij}\mathbf{e}_i\mathbf{e}_j] \\
&= -\mathbf{e}_z \cdot [(r\mathbf{e}_r + z\mathbf{e}_z) \times (P\mathbf{e}_r + \tau_{rj}\mathbf{e}_j] \\
&= -\mathbf{e}_z \cdot (zP\mathbf{e}_\theta + \epsilon_{rjk}r\tau_{rj}\mathbf{e}_k + \epsilon_{zjk}z\tau_{rj}\mathbf{e}_k) \\
&= -zP\underbrace{(\mathbf{e}_z \cdot \mathbf{e}_\theta)}_{0} - \underbrace{\epsilon_{rjz}r\tau_{rj}}_{j=\theta} - \underbrace{\epsilon_{zjz}z\tau_{rj}}_{0} \\
&= -r\tau_{r\theta}. \tag{7.6-4}
\end{aligned}$$

The term A in Eq. (7.6-3) represents the lateral surface of the outer cylinder. Therefore, $dA = R\,d\theta dz$ and Eq. (7.6-3) becomes

$$T_z = -\int_0^L \int_0^{2\pi} (r\tau_{r\theta})|_{r=R}\, R\, d\theta dz = -2\pi LR\,(r\tau_{r\theta})|_{r=R}, \tag{7.6-5}$$

where L is the length of the cylinder. Substitution of Eq. (7.6-2) into Eq. (7.6-5) yields

$$T_z = -2\pi L C_1 \Rightarrow C_1 = -\frac{T_z}{2\pi L}. \tag{7.6-6}$$

Therefore, shear stress distribution takes the following form:

$$\tau_{r\theta} = -\frac{T_z}{2\pi L}\frac{1}{r^2}. \tag{7.6-7}$$

For a Bingham fluid, the non-Newtonian viscosity is given by Eq. (7.1-7). The use of Eq. (7.2-25) gives

$$\frac{1}{2}(\boldsymbol{\tau}:\boldsymbol{\tau}) = \frac{1}{2}(\tau_{r\theta}^2 + \tau_{\theta r}^2) = \tau_{r\theta}^2. \tag{7.6-8}$$

Since $\tau_{r\theta}$ is negative, from Eq. (7.1-5)

$$\tau = |\tau_{r\theta}| = -\tau_{r\theta}. \tag{7.6-9}$$

Let r_o be the radial distance at which $-\tau_{r\theta} = \tau_o$. Thus, from Eq. (7.6-7)

$$-\tau_o = -\frac{T_z}{2\pi L}\frac{1}{r_o^2} \quad \Rightarrow \quad r_o = \sqrt{\frac{T_z}{2\pi L \tau_o}}. \tag{7.6-10}$$

Once T_z and τ_o are specified, the value of r_o, calculated from Eq. (7.6-10), may be less than κR, between κR and R, and greater than R.

Case (i) $r_o \leq \kappa R$:
In this case, there will be no flow.

Case (ii) $\kappa R < r_o \leq R$:
The magnitude of the shear rate is determined from Eq. (B) in Table 7.1 as

$$\dot{\gamma} = \left| r\frac{d}{dr}\left(\frac{v_\theta}{r}\right) \right|. \tag{7.6-11}$$

Since v_θ/r increases with increasing r, Eq. (7.6-11) becomes

$$\dot{\gamma} = r\frac{d}{dr}\left(\frac{v_\theta}{r}\right). \tag{7.6-12}$$

In this case, the flow region is divided into two zones, namely, region I ($r_o \leq r \leq R$), in which $\tau_{r\theta} \leq \tau_o$, and region II ($\kappa R \leq r \leq r_o$), in which $\tau_{r\theta} \geq \tau_o$.

- **Region I** $(\tau_{r\theta} \leq \tau_o)$

Since $\eta = \infty$, rearrangement of Eq. (7.6-1) in the form

$$r \frac{d}{dr}\left(\frac{v_\theta^{\mathrm{I}}}{r}\right) = -\frac{1}{\eta}\tau_{r\theta},$$ (7.6-13)

implies that

$$\frac{d}{dr}\left(\frac{v_\theta^{\mathrm{I}}}{r}\right) = 0 \Rightarrow v_\theta^{\mathrm{I}} = C_2 r.$$ (7.6-14)

The use of the boundary condition

$$\text{at } r = R \qquad v_\theta^{\mathrm{I}} = \Omega R,$$ (7.6-15)

gives the velocity distribution as

$$\frac{v_\theta^{\mathrm{I}}}{r} = \Omega.$$ (7.6-16)

In this geometric configuration, note that $dv_\theta/dr \neq 0$ when $\tau_{r\theta} \leq \tau_o$.

- **Region II** $(\tau_{r\theta} \geq \tau_o)$

The use of Eq. (7.6-12) in Eq. (7.1-7b) gives

$$\eta = \mu_o + \frac{\tau_o}{\dot{\gamma}} = \mu_o + \frac{\tau_o}{r\dfrac{d}{dr}\left(\dfrac{v_\theta^{\mathrm{II}}}{r}\right)}.$$ (7.6-17)

Substitution of Eq. (7.6-17) into Eq. (7.6-1) yields

$$\tau_{r\theta} = -\left[\mu_o + \frac{\tau_o}{r\dfrac{d}{dr}\left(\dfrac{v_\theta^{\mathrm{II}}}{r}\right)}\right] r \frac{d}{dr}\left(\frac{v_\theta^{\mathrm{II}}}{r}\right)$$

$$= -\mu_o r \frac{d}{dr}\left(\frac{v_\theta^{\mathrm{II}}}{r}\right) - \tau_o.$$ (7.6-18)

Equating Eq. (7.6-7) to Eq. (7.6-18) and then rearranging lead to

$$\frac{d}{dr}\left(\frac{v_\theta^{\mathrm{II}}}{r}\right) = \frac{T_z}{2\pi\mu_o L}\frac{1}{r^3} - \frac{\tau_o}{\mu_o r}.$$ (7.6-19)

Integration of Eq. (7.6-19) results in

$$\frac{v_\theta^{\mathrm{II}}}{r} = -\frac{T_z}{4\pi\mu_o L}\frac{1}{r^2} - \frac{\tau_o}{\mu_o}\ln r + C_3.$$ (7.6-20)

The use of the boundary condition

$$\text{at } r = r_o \qquad v_\theta^{\text{II}} = v_\theta^{\text{I}}, \tag{7.6-21}$$

gives the velocity distribution as

$$\frac{v_\theta^{\text{II}}}{r} = \Omega + \frac{T_z}{4\pi\mu_o L r_o^2} \left[1 - \left(\frac{r_o}{r}\right)^2 \right] - \frac{\tau_o}{\mu_o} \ln\left(\frac{r}{r_o}\right). \tag{7.6-22}$$

Case (iii) $r_o \geq R$:
In this case, region I of case (ii) does not exist. Taking $r_o = R$ in Eq. (7.6-22) gives the velocity distribution as

$$\frac{v_\theta}{r} = \Omega + \frac{T_z}{4\pi\mu_o L R^2} \left[1 - \left(\frac{R}{r}\right)^2 \right] - \frac{\tau_o}{\mu_o} \ln\left(\frac{r}{R}\right). \tag{7.6-23}$$

Application of the boundary condition

$$\text{at } r = \kappa R \quad v_\theta = 0, \tag{7.6-24}$$

gives the relationship between Ω and T_z in the following form:

$$\Omega = \frac{T_z}{4\pi\mu_o L R^2} \left(\frac{1}{\kappa^2} - 1 \right) + \frac{\tau_o}{\mu_o} \ln\kappa. \tag{7.6-25}$$

7.7 Generalized Couette Flow of a Herschel–Bulkley Fluid Between Parallel Plates

For steady, laminar flow of an incompressible Herschel–Bulkley fluid in the flow geometry shown in Figure 7.5, the postulates that $v_z = v_z(x)$ and $v_x = v_y = 0$ simplify the z-component of the equation of motion as

$$0 = \frac{\mathcal{P}_o - \mathcal{P}_L}{L} - \frac{d\tau_{xz}}{dx}, \tag{7.7-1}$$

where $\mathcal{P}$ is the modified pressure defined by

$$\mathcal{P} = P + \rho g x. \tag{7.7-2}$$

Integration of Eq. (7.7-1) gives

$$\tau_{xz} = \left(\frac{\mathcal{P}_o - \mathcal{P}_L}{L}\right) x + C, \tag{7.7-3}$$

where C is the integration constant. The use of the boundary condition

$$\tau_{xz} = 0 \quad \text{at } x = \lambda H, \tag{7.7-4}$$

gives the shear stress distribution as

$$\tau_{xz} = \left(\frac{\mathcal{P}_o - \mathcal{P}_L}{L}\right) H \left(\frac{x}{H} - \lambda\right), \tag{7.7-5}$$

in which the value of λ will be determined later.

For a flow between parallel plates,

$$\tau = |\tau_{xz}| \quad \text{and} \quad \dot{\gamma} = \left|\frac{dv_z}{dx}\right|. \tag{7.7-6}$$

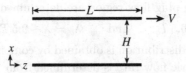

Figure 7.5 Flow of a Herschel–Bulkley fluid between parallel plates.

Therefore, Eqs. (7.1-9a) and (7.1-9b) simplify to

$$\frac{dv_z}{dx} = 0 \quad \text{when } |\tau_{xz}| \leq \tau_0, \tag{7.7-7a}$$

$$\tau_{xz} = \pm \tau_0 - m \left| \frac{dv_z}{dx} \right|^{n-1} \frac{dv_z}{dx} \quad \text{when } |\tau_{xz}| \geq \tau_0. \tag{7.7-7b}$$

The positive sign in Eq. (7.7-7b) is to be used when τ_{xz} is positive and the negative sign is to be used when τ_{xz} is negative.

Introduction of the following dimensionless quantities:

$$\phi = \frac{v_z}{V} \qquad \xi = \frac{x}{H} \qquad T = \frac{2\tau_{xz}}{\left(\dfrac{P_o - P_L}{L}\right) H}$$

$$\Lambda = \frac{(P_o - P_L)H}{mL} \left(\frac{H}{V}\right)^n, \tag{7.7-8}$$

reduces Eqs. (7.7-5), (7.7-7a), and (7.7-7b) to

$$T = 2 \left(\xi - \lambda\right), \tag{7.7-9}$$

and

$$\frac{d\phi}{d\xi} = 0 \quad \text{when } |T| \leq T_o, \tag{7.7-10a}$$

$$T = \pm T_o - \frac{2}{\Lambda} \left| \frac{d\phi}{d\xi} \right|^{n-1} \frac{d\phi}{d\xi} \quad \text{when } |T| \geq T_o, \tag{7.7-10b}$$

where T_o is defined by

$$T_o = \frac{2\tau_o}{\left(\dfrac{P_o - P_L}{L}\right) H}. \tag{7.7-11}$$

Equation (7.7-10a) indicates that the plug flow region exists when $-T_o \leq T \leq T_o$. Using

$$\xi = \begin{cases} \lambda_i & \text{at } T = -T_o, \\ \lambda_o & \text{at } T = T_o, \end{cases} \tag{7.7-12}$$

the boundaries of the plug flow region are determined as

$$\lambda_i = \lambda - 0.5\,T_o \qquad \text{and} \qquad \lambda_o = \lambda + 0.5\,T_o. \qquad (7.7\text{-}13)$$

Once the velocity distribution is obtained by combining Eqs. (7.7-9) and (7.7-10), the volumetric flow rate is determined from

$$\mathcal{Q} = \int_0^W \int_0^H v_z \, dx \, dy, \qquad (7.7\text{-}14)$$

where W is the width of the plate. Integration of Eq. (7.7-14) by parts leads to

$$\mathcal{Q} = W \left[H\, v_z|_{x=H} - \int_0^H x \left(\frac{dv_z}{dx} \right) dx \right]. \qquad (7.7\text{-}15)$$

In terms of the dimensionless quantities, Eq. (7.7-15) takes the form

$$\overline{\mathcal{Q}} = \pm 1 - \left[\int_0^{\lambda_i} \xi \left(\frac{d\phi}{d\xi} \right) d\xi + \int_{\lambda_o}^1 \xi \left(\frac{d\phi}{d\xi} \right) d\xi \right], \qquad (7.7\text{-}16)$$

where the dimensionless volumetric flow rate, $\overline{\mathcal{Q}}$, is defined by

$$\overline{\mathcal{Q}} = \frac{\mathcal{Q}}{WBV}. \qquad (7.7\text{-}17)$$

The positive sign in Eq. (7.7-16) is to be used when the upper plate moves in the positive z-direction, and the negative sign is to be used when the upper plate moves in the negative z-direction.

The volumetric flow rate is dependent on the imposed pressure gradient, $(\mathcal{P}_o - \mathcal{P}_L)/L$, as well as the magnitude and direction of the upper plate velocity, V. When the upper plate moves in the direction of the decreasing pressure gradient, the location of the maximum velocity depends on the relative magnitudes of the plate velocity and the pressure gradient. Flow type A considers the case in which the maximum velocity occurs in the plug flow region. In the case of flow type B, the maximum velocity occurs at the upper plate. Flow type C, on the other hand, examines the case when the plate moves in the direction opposite to the direction of the decreasing pressure gradient. These three cases are depicted in Figure 7.6. In each flow type, the flow domain is divided into three regions.

Table 7.2 shows the values of the dimensionless shear stress, T, expression given by Eq. (7.7-9) in different regions for each flow type.

Once the expressions in Table 7.2 are combined with Eq. (7.7-9) and integrated with the following boundary conditions:

$$\text{at } \xi = 0 \qquad \phi^I = 0, \qquad (7.7\text{-}18)$$

$$\text{at } \xi = 1 \qquad \phi^{III} = \pm 1, \qquad (7.7\text{-}19)$$

the resulting velocity distributions are shown in Table 7.3 in terms of λ, the dimensionless distance at which shear stress is zero, and $s = 1/n$.

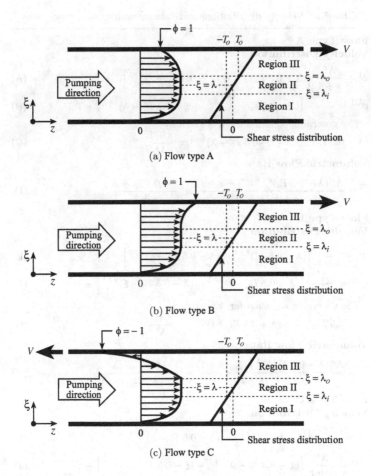

(a) Flow type A

(b) Flow type B

(c) Flow type C

Figure 7.6   Velocity and shear stress distributions for three different types of flow.

Table 7.2   Values of the dimensionless shear stress, T, in regions I, II, and III depending on the flow type.

Flow type	Region I $(0 \leq \xi \leq \lambda_i)$	Region II $(\lambda_i \leq \xi \leq \lambda_o)$	Region III $(\lambda_o \leq \xi \leq 1)$
A	$T = -T_o - \dfrac{2}{\Lambda} \left(\dfrac{d\phi^{\mathrm{I}}}{d\xi} \right)^n$	$\dfrac{d\phi^{\mathrm{II}}}{d\xi} = 0$	$T = T_o + \dfrac{2}{\Lambda} \left(-\dfrac{d\phi^{\mathrm{III}}}{d\xi} \right)^n$
B	$T = -T_o - \dfrac{2}{\Lambda} \left(\dfrac{d\phi^{\mathrm{I}}}{d\xi} \right)^n$	$\dfrac{d\phi^{\mathrm{II}}}{d\xi} = 0$	$T = T_o - \dfrac{2}{\Lambda} \left(\dfrac{d\phi^{\mathrm{III}}}{d\xi} \right)^n$
C	$T = -T_o - \dfrac{2}{\Lambda} \left(\dfrac{d\phi^{\mathrm{I}}}{d\xi} \right)^n$	$\dfrac{d\phi^{\mathrm{II}}}{d\xi} = 0$	$T = T_o + \dfrac{2}{\Lambda} \left(-\dfrac{d\phi^{\mathrm{III}}}{d\xi} \right)^n$

Table 7.3   Velocity distributions and volumetric flow rate expressions.

Flow Type A
Velocity Distribution

$$\phi^{\mathrm{I}} = \frac{\Lambda^s}{s+1} \left[(\lambda - 0.5\,T_o)^{s+1} - (\lambda - 0.5\,T_o - \xi)^{s+1} \right] \tag{A}$$

$$\phi^{\mathrm{III}} = \frac{\Lambda^s}{s+1} \left[(1 - 0.5\,T_o - \lambda)^{s+1} - (\xi - 0.5\,T_o - \lambda)^{s+1} \right] + 1 \tag{B}$$

The governing equation for λ is given by

$$(\lambda - 0.5\,T_o)^{s+1} - (1 - 0.5\,T_o - \lambda)^{s+1} = \frac{s+1}{\Lambda^s} \tag{C}$$

Volumetric Flow Rate

$$\overline{Q} = \frac{\Lambda^s\,(\lambda - 0.5\,T_o)^{s+1}\,(T_o + s + 1)}{(s+1)\,(s+2)} + \frac{1 - 0.5\,T_o - \lambda}{s+2} \tag{D}$$

Flow Type B
Velocity Distribution

$$\phi^{\mathrm{I}} = \frac{\Lambda^s}{s+1} \left[(\lambda - 0.5\,T_o)^{s+1} - (\lambda - 0.5\,T_o - \xi)^{s+1} \right] \tag{E}$$

$$\phi^{\mathrm{III}} = \frac{\Lambda^s}{s+1} \left[(\lambda + 0.5\,T_o - 1)^{s+1} - (\lambda + 0.5\,T_o - \xi)^{s+1} \right] + 1 \tag{F}$$

The governing equation for λ is given by

$$(\lambda - 0.5\,T_o)^{s+1} - (\lambda + 0.5\,T_o - 1)^{s+1} = \frac{s+1}{\Lambda^s} \tag{G}$$

Volumetric Flow Rate

$$\overline{Q} = \frac{\Lambda^s\,(\lambda - 0.5\,T_o)^{s+1}\,(T_o + s + 1)}{(s+1)\,(s+2)} + \frac{1 - 0.5\,T_o - \lambda}{s+2} \tag{H}$$

Flow Type C
Velocity Distribution

$$\phi^{\mathrm{I}} = \frac{\Lambda^s}{s+1} \left[(\lambda - 0.5\,T_o)^{s+1} - (\lambda - 0.5\,T_o - \xi)^{s+1} \right] \tag{I}$$

$$\phi^{\mathrm{III}} = \frac{\Lambda^s}{s+1} \left[(1 - 0.5\,T_o - \lambda)^{s+1} - (\xi - 0.5\,T_o - \lambda)^{s+1} \right] - 1 \tag{J}$$

The governing equation for λ is given by

$$(1 - 0.5\,T_o - \lambda)^{s+1} - (\lambda - 0.5\,T_o)^{s+1} = \frac{s+1}{\Lambda^s} \tag{K}$$

Volumetric Flow Rate

$$\overline{Q} = \frac{\Lambda^s\,(\lambda - 0.5\,T_o)^{s+1}\,(T_o + s + 1)}{(s+1)\,(s+2)} + \frac{\lambda + 0.5\,T_o - 1}{s+2} \tag{L}$$

Chapter 8

Creeping Flow

8.1 The Criterion for Creeping Flow Assumption

The Navier–Stokes equation, Eq. (1.2-22), under steady-state flow conditions is expressed as

$$\underbrace{\rho \mathbf{v} \cdot \nabla \mathbf{v}}_{\substack{\text{Inertial} \\ \text{term}}} = -\nabla P + \underbrace{\mu \nabla^2 \mathbf{v}}_{\substack{\text{Viscous} \\ \text{term}}} + \rho \mathbf{g}. \qquad (8.1\text{-}1)$$

While the inertial terms arise from the fluid acceleration, viscous terms arise from the shear stresses. Simplification of the Navier–Stokes equation is dictated by the magnitudes of the inertial and viscous terms.[1]

The order of magnitude of the inertial term is

$$\rho \mathbf{v} \cdot \nabla \mathbf{v} = O\left(\frac{\rho v_{\text{ch}}^2}{L_{\text{ch,i}}}\right), \qquad (8.1\text{-}2)$$

where v_{ch} is the characteristic velocity, and $L_{\text{ch,i}}$ is the "inertial length" representing the characteristic distance along the streamline over which significant variations in the velocity take place.

The order of magnitude of the viscous term is

$$\mu \nabla^2 \mathbf{v} = O\left(\frac{\mu v_{\text{ch}}}{L_{\text{ch,v}}^2}\right), \qquad (8.1\text{-}3)$$

where $L_{\text{ch,v}}$ is the characteristic "viscous length".

---

[1]For a more thorough discussion on the subject, see Whitaker (1988).

Therefore, the ratio of the inertial to viscous terms is given by

$$\frac{\text{Inertial term}}{\text{Viscous term}} = \frac{L_{ch,v} v_{ch} \rho}{\mu} \frac{L_{ch,v}}{L_{ch,i}}. \tag{8.1-4}$$

The Reynolds number, Re, is defined in terms of the viscous length in the form

$$\text{Re} = \frac{L_{ch,v} v_{ch} \rho}{\mu}, \tag{8.1-5}$$

so that Eq. (8.1-4) is expressed as

$$\frac{\text{Inertial terms}}{\text{Viscous terms}} = \text{Re}\left(\frac{L_{ch,v}}{L_{ch,i}}\right). \tag{8.1-6}$$

A flow is said to be *creeping* when the viscous term dominates over the inertial term, i.e., inertial term $\gg$ viscous term. Therefore,

$$\text{Re}\left(\frac{L_{ch,v}}{L_{ch,i}}\right) \ll 1 \quad \text{criterion for creeping flow.} \tag{8.1-7}$$

This is the case when the Reynolds number is small. Note that the criterion of Re $\ll$ 1 is satisfied not only for low characteristic velocities but also for fluids having high viscosity and for cases in which $L_{ch,v} \ll 1$. For a creeping flow under steady conditions, the left side of the Navier–Stokes equation becomes zero, i.e.,

$$0 = -\nabla P + \mu \nabla^2 \mathbf{v} + \rho \mathbf{g}. \tag{8.1-8}$$

8.2   Flow Between Stationary Concentric Spheres

Consider steady, creeping flow of an incompressible Newtonian fluid between two concentric spheres as shown in Figure 8.1(a).

From Figure 8.1(b), the components of the gravity vector are

$$g_r = -g\cos\theta \qquad g_\theta = g\sin\theta \qquad g_\phi = 0. \tag{8.2-1}$$

For a one-dimensional flow under steady conditions, the velocity components are

$$v_r = v_\phi = 0 \quad \text{and} \quad v_\theta = v_\theta(r, \theta, \phi). \tag{8.2-2}$$

Using the condition of angular symmetry

$$v_\theta(r, \theta, \phi) = v_\theta(r, \theta, \phi + \pi), \tag{8.2-3}$$

one concludes that $v_\theta = v_\theta(r, \theta)$.

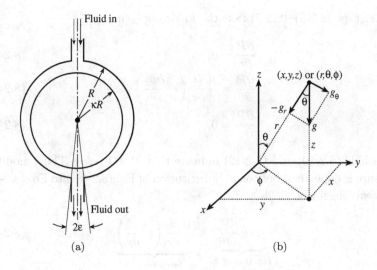

Figure 8.1   Flow between concentric spheres.

The equation of continuity, Eq. (C) in Table 1.1, simplifies to

$$\frac{\partial}{\partial\theta}(v_\theta \sin\theta) = 0 \quad \Rightarrow \quad v_\theta \sin\theta = f(r) \quad \Rightarrow \quad v_\theta = \frac{f(r)}{\sin\theta}. \qquad (8.2\text{-}4)$$

The components of the equation of motion, Eqs. (D), (E), and (F) in Table 1.7, become

$$r\text{-component} \quad 0 = -\frac{\partial P}{\partial r} - \rho g \cos\theta, \qquad (8.2\text{-}5)$$

$$\theta\text{-component} \quad 0 = -\frac{1}{r}\frac{\partial P}{\partial\theta} + \frac{\mu}{r^2}\frac{\partial}{\partial r}\left(r^2\frac{\partial v_\theta}{\partial r}\right) + \rho g \sin\theta, \qquad (8.2\text{-}6)$$

$$\phi\text{-component} \quad 0 = \frac{\partial P}{\partial\phi}. \qquad (8.2\text{-}7)$$

Since

$$z = r \cos\theta, \qquad (8.2\text{-}8)$$

the modified pressure is defined as

$$\mathcal{P} = P + \rho g r \cos\theta. \qquad (8.2\text{-}9)$$

Hence, Eqs. (8.2-5)–(8.2-7) take the following form:

$$\frac{\partial P}{\partial r} = 0, \tag{8.2-10}$$

$$\frac{\partial P}{\partial \theta} = \frac{\mu}{r} \frac{\partial}{\partial r} \left(r^2 \frac{\partial v_\theta}{\partial r} \right), \tag{8.2-11}$$

$$\frac{\partial P}{\partial \phi} = 0. \tag{8.2-12}$$

Equations (8.2-10) and (8.2-12) indicate that $P \neq P(r, \phi)$. Thus, modified pressure is dependent only on θ. Substitution of Eq. (8.2-4) into Eq. (8.2-11) and rearrangement yield

$$\underbrace{\sin\theta \frac{dP}{d\theta}}_{\text{function of } \theta} = \underbrace{\frac{\mu}{r} \frac{d}{dr} \left(r^2 \frac{df}{dr} \right)}_{\text{function of } r}. \tag{8.2-13}$$

The term on the right-hand side of Eq. (8.2-13) is a function only of r, while the term on the left-hand side depends only on θ. This is possible if both sides are equal to a constant, i.e.,

$$\sin\theta \frac{dP}{d\theta} = A = \text{constant.} \tag{8.2-14}$$

Rearrangement of Eq. (8.2-14) as

$$\int_{P_1}^{P_2} dP = A \int_{\epsilon}^{\pi-\epsilon} \frac{d\theta}{\sin\theta}, \tag{8.2-15}$$

and integration[2] result in

$$A = -\frac{P_1 - P_2}{\Xi}, \tag{8.2-16}$$

where P_1 and P_2 are modified pressures at the inlet and outlet, respectively, and Ξ is defined by

$$\Xi = \ln\left(\frac{1 + \cos\epsilon}{1 - \cos\epsilon} \right). \tag{8.2-17}$$

---

[2]Note that

$$\int \frac{d\theta}{\sin\theta} = \ln[\tan(\theta/2)] = \frac{1}{2} \ln\left(\frac{1 - \cos\theta}{1 + \cos\theta} \right)$$

and $\cos(\pi - \theta) = -\cos\theta$.

Substitution of Eq. (8.2-16) into Eq. (8.2-13) results in

$$\frac{\mu}{r}\frac{d}{dr}\left(r^2\frac{df}{dr}\right) = -\frac{(\mathcal{P}_1 - \mathcal{P}_2)}{\Xi}. \qquad (8.2\text{-}18)$$

Integration of Eq. (8.2-18) twice gives

$$f = -\frac{(\mathcal{P}_1 - \mathcal{P}_2)}{2\mu\Xi}r - \frac{C_1}{r} + C_2. \qquad (8.2\text{-}19)$$

The use of the boundary conditions

$$\text{at } r = \kappa R \quad v_\theta = 0 \quad \Rightarrow \quad f = 0, \qquad (8.2\text{-}20)$$
$$\text{at } r = R \quad v_\theta = 0 \quad \Rightarrow \quad f = 0, \qquad (8.2\text{-}21)$$

gives the function f as

$$f = \frac{(\mathcal{P}_1 - \mathcal{P}_2)R}{2\mu\Xi}\left[1 - \frac{r}{R} + \kappa\left(1 - \frac{R}{r}\right)\right]. \qquad (8.2\text{-}22)$$

Substitution of Eq. (8.2-22) into Eq. (8.2-4) gives the velocity distribution as

$$v_\theta = \frac{(\mathcal{P}_1 - \mathcal{P}_2)R}{2\mu\Xi\sin\theta}\left[1 - \frac{r}{R} + \kappa\left(1 - \frac{R}{r}\right)\right]. \qquad (8.2\text{-}23)$$

From Table 1.4, the only non-zero shear stress component is

$$\tau_{r\theta} = -\mu r \frac{\partial}{\partial r}\left(\frac{v_\theta}{r}\right). \qquad (8.2\text{-}24)$$

Substitution of Eq. (8.2-23) into Eq. (8.2-24) yields

$$\tau_{r\theta} = \frac{(\mathcal{P}_1 - \mathcal{P}_2)R}{2\Xi\sin\theta}\left(\frac{\kappa + 1}{r} - \frac{2\kappa R}{r^2}\right). \qquad (8.2\text{-}25)$$

For one-dimensional flow in Cartesian and cylindrical coordinate systems in which τ_{xz} and τ_{rz} are the only non-vanishing shear stress components, respectively, it is obvious that the shear stress becomes zero at the point where velocity reaches its maximum value. This leads to a misconception that shear stress is "always" zero at the point of maximum velocity. As will be shown in this problem, however, this is not always true. From Eq. (8.2-23), the point at which maximum velocity occurs is

$$\frac{dv_\theta}{dr} = 0 \quad \Rightarrow \quad r = R\sqrt{\kappa}. \qquad (8.2\text{-}26)$$

On the other hand, the point at which the shear stress is zero is calculated from Eq. (8.2-25) as

$$\tau_{r\theta} = 0 \qquad \Rightarrow \qquad r = \frac{2\kappa R}{1 + \kappa}, \tag{8.2-27}$$

which is completely different from Eq. (8.2-26).

The volumetric flow rate can be determined by integrating the velocity distribution over the cross-sectional area, i.e.,

$$Q = \int_0^{2\pi} \int_{\kappa R}^{R} v_\theta r \sin\theta \, dr \, d\phi = 2\pi \int_{\kappa R}^{R} f(r) r \, dr. \tag{8.2-28}$$

In terms of the dimensionless distance ξ, defined by

$$\xi = \frac{r}{R}, \tag{8.2-29}$$

substitution of Eq. (8.2-22) into Eq. (8.2-28) leads to

$$Q = \frac{(\mathcal{P}_1 - \mathcal{P}_2)\pi R^3}{\mu\Xi} \int_\kappa^1 \left[\xi - \xi^2 + \kappa(\xi - 1) \right] d\xi. \tag{8.2-30}$$

Integration of Eq. (8.2-30) gives the volumetric flow rate as

$$Q = \frac{(\mathcal{P}_1 - \mathcal{P}_2)\pi R^3 (1 - \kappa)^3}{6\mu\Xi}. \tag{8.2-31}$$

8.2.1 Pressure distribution

Substitution of Eqs. (8.2-9) and (8.2-16) into Eq. (8.2-14) and rearrangement result in

$$\frac{\partial P}{\partial \theta} = -\frac{(\mathcal{P}_1 - \mathcal{P}_2)}{\Xi} \frac{1}{\sin\theta} + \rho g r \sin\theta. \tag{8.2-32}$$

Integration of Eq. (8.2-32) gives the pressure distribution as

$$P = -\frac{(\mathcal{P}_1 - \mathcal{P}_2)}{2\Xi} \ln\left(\frac{1 - \cos\theta}{1 + \cos\theta}\right) - \rho g r \cos\theta + u(r). \tag{8.2-33}$$

8.2.2 Calculation of drag force

The drag force exerted by the fluid on the surfaces of the spheres is given by Eq. (4.1-15), in which $\boldsymbol{\lambda} = \mathbf{e}_z$. The use of Eq. (A.6-17) in Appendix A gives

$$\boldsymbol{\lambda} = \mathbf{e}_z = \cos\theta \, \mathbf{e}_r - \sin\theta \, \mathbf{e}_\theta, \tag{8.2-34}$$

and the unit vector directed from the fluid to the surface of the sphere is given by

$$
\mathbf{n} = \begin{cases} -\mathbf{e}_r & \text{at } r = \kappa R \\ \mathbf{e}_r & \text{at } r = R. \end{cases} \tag{8.2-35}
$$

Note that the differential area is $dA = r^2 \sin\theta \, d\theta \, d\phi$. Therefore, Eq. (4.1-15) simplifies to[3]

$$
F_D = -\underbrace{\int_0^{2\pi} \int_\epsilon^{\pi-\epsilon} \cos\theta \, P|_{r=\kappa R} \, \kappa^2 R^2 \sin\theta \, d\theta \, d\phi}_{\text{I}}
$$

$$
+ \underbrace{\int_0^{2\pi} \int_\epsilon^{\pi-\epsilon} \cos\theta \, P|_{r=R} \, R^2 \sin\theta \, d\theta \, d\phi}_{\text{II}}
$$

$$
- \underbrace{\int_0^{2\pi} \int_\epsilon^{\pi-\epsilon} \sin\theta \, \mathbf{e}_r \mathbf{e}_\theta{:}\tau|_{r=\kappa R} \, \kappa^2 R^2 \sin\theta \, d\theta \, d\phi}_{\text{III}}
$$

$$
+ \underbrace{\int_0^{2\pi} \int_\epsilon^{\pi-\epsilon} \sin\theta \, \mathbf{e}_r \mathbf{e}_\theta{:}\tau|_{r=R} \, R^2 \sin\theta \, d\theta \, d\phi}_{\text{IV}}. \tag{8.2-36}
$$

- **Evaluation of the term I**

  Substitution of Eq. (8.2-33) gives

$$
I = 2\pi \left[\frac{(\mathcal{P}_1 - \mathcal{P}_2)\kappa^2 R^2}{2\Xi} \int_\epsilon^{\pi-\epsilon} \ln\left(\frac{1-\cos\theta}{1+\cos\theta}\right) \sin\theta \cos\theta \, d\theta \right.
$$

$$
\left. + \rho g \kappa R \underbrace{\int_\epsilon^{\pi-\epsilon} \sin\theta \cos^2\theta \, d\theta}_{0} - u(r) \underbrace{\int_\epsilon^{\pi-\epsilon} \sin\theta \cos\theta \, d\theta}_{0} \right]. \tag{8.2-37}
$$

Hence,

$$
I = \frac{\pi(\mathcal{P}_1 - \mathcal{P}_2)\kappa^2 R^2}{\Xi} \int_\epsilon^{\pi-\epsilon} \ln\left(\frac{1-\cos\theta}{1+\cos\theta}\right) \sin\theta \cos\theta \, d\theta. \tag{8.2-38}
$$

---

[3] Note that

$$
\boldsymbol{\lambda} \cdot (\mathbf{n} \cdot \boldsymbol{\tau}) = \boldsymbol{\lambda}\mathbf{n} : \boldsymbol{\tau}.
$$

- **Evaluation of the term II**

  Substitution of Eq. (8.2-33) gives

$$\text{II} = 2\pi \left[-\frac{(\mathcal{P}_1 - \mathcal{P}_2)R^2}{2\Xi} \int_\epsilon^{\pi-\epsilon} \ln\left(\frac{1-\cos\theta}{1+\cos\theta}\right) \sin\theta \cos\theta \, d\theta \right.$$

$$\left. -\rho g R \underbrace{\int_\epsilon^{\pi-\epsilon} \sin\theta \cos^2\theta \, d\theta}_{0} + u(r) \underbrace{\int_\epsilon^{\pi-\epsilon} \sin\theta \cos\theta \, d\theta}_{0} \right].$$

$$(8.2\text{-}39)$$

Hence,

$$\text{II} = -\frac{\pi(\mathcal{P}_1 - \mathcal{P}_2)R^2}{\Xi} \int_\epsilon^{\pi-\epsilon} \ln\left(\frac{1-\cos\theta}{1+\cos\theta}\right) \sin\theta \cos\theta \, d\theta. \quad (8.2\text{-}40)$$

- **Evaluation of the term III**

  Simplification gives

$$\text{III} = -2\pi\kappa^2 R^2 \int_\epsilon^{\pi-\epsilon} \sin\theta \left. \tau_{r\theta}\right|_{r=\kappa R} \sin\theta \, d\theta. \quad (8.2\text{-}41)$$

Substitution of Eq. (8.2-25) into Eq. (8.2-41) yields

$$\text{III} = \frac{\pi(\mathcal{P}_1 - \mathcal{P}_2)\kappa R^2(1-\kappa)}{\Xi} \underbrace{\int_\epsilon^{\pi-\epsilon} \sin\theta \, d\theta}_{2\cos\epsilon}$$

$$= \frac{2\pi(\mathcal{P}_1 - \mathcal{P}_2)\kappa R^2(1-\kappa)\cos\epsilon}{\Xi}. \quad (8.2\text{-}42)$$

- **Evaluation of the term IV**

  Simplification gives

$$\text{IV} = 2\pi R^2 \int_\epsilon^{\pi-\epsilon} \sin\theta \left. \tau_{r\theta}\right|_{r=R} \sin\theta \, d\theta. \quad (8.2\text{-}43)$$

Substitution of Eq. (8.2-25) into Eq. (8.2-43) yields

$$\text{IV} = \frac{\pi(\mathcal{P}_1 - \mathcal{P}_2)R^2(1-\kappa)}{\Xi} \int_\epsilon^{\pi-\epsilon} \sin\theta \, d\theta$$

$$= \frac{2\pi(\mathcal{P}_1 - \mathcal{P}_2)R^2(1-\kappa)\cos\epsilon}{\Xi}. \quad (8.2\text{-}44)$$

Therefore, the form drag is

$$I + II = \frac{\pi R^2 (\mathcal{P}_1 - \mathcal{P}_2)(1 - \kappa^2)}{\Xi} \int_\epsilon^{\pi - \epsilon} \ln \left(\frac{1 + \cos \theta}{1 - \cos \theta} \right) \sin \theta \cos \theta \, d\theta.$$

$$(8.2\text{-}45)$$

The integral in Eq. (8.2-45) must be evaluated numerically. On the other hand, the friction drag is

$$III + IV = \pi R^2 (1 - \kappa^2) \left(\frac{2 \cos \epsilon}{\Xi} \right) (\mathcal{P}_1 - \mathcal{P}_2). \qquad (8.2\text{-}46)$$

8.3 Rotating Sphere in an Infinite Medium

A solid sphere of radius R is immersed in a large volume of a Newtonian fluid. While it slowly rotates at an angular velocity Ω along the z-axis, as shown in Figure 8.2, the fluid far from the sphere is stagnant.

Mathematical Preliminary. The angular velocity vector of a rigid object rotating about the z-axis is given by

$$\boldsymbol{\Omega} = \Omega \, \mathbf{e}_z. \qquad (8.3\text{-}1)$$

Substitution of Eq. (A.6-17) in Appendix A into Eq. (8.3-1) gives

$$\boldsymbol{\Omega} = \Omega (\cos \theta \, \mathbf{e}_r - \sin \theta \, \mathbf{e}_\theta). \qquad (8.3\text{-}2)$$

At any point in the rotating object, the linear velocity is given by

$$\mathbf{v} = \boldsymbol{\Omega} \times \mathbf{r}, \qquad (8.3\text{-}3)$$

where $\mathbf{r}$ is the position vector to that point.

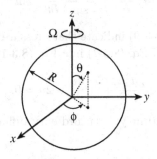

Figure 8.2   Sphere rotating in an infinite fluid.

The position vector in the spherical coordinate system is given by Eq. (4.2-28) as $\mathbf{r} = r\mathbf{e}_r$. Therefore, Eq. (8.3-3) is expressed as

$$\mathbf{v} = \Omega r \cos\theta \underbrace{(\mathbf{e}_r \times \mathbf{e}_r)}_{0} - \Omega r \sin\theta (\mathbf{e}_\theta \times \mathbf{e}_r)$$

$$= -\Omega r \sin\theta \underbrace{\epsilon_{\theta r\phi}}_{-1} \mathbf{e}_\phi = \Omega r \sin\theta\, \mathbf{e}_\phi. \qquad (8.3\text{-}4)$$

Rotational symmetry around the z-axis implies that the components of velocity and pressure are independent of ϕ. The boundary conditions are

$$\text{at } r = R \quad v_r = v_\theta = 0 \ \& \ v_\phi = \Omega R \sin\theta, \qquad (8.3\text{-}5)$$

$$\text{as } r \to \infty \quad v_r = v_\theta = v_\phi = 0. \qquad (8.3\text{-}6)$$

If we postulate

$$v_r = v_\theta = 0 \qquad \text{and} \qquad v_\phi = u(r)\sin\theta, \qquad (8.3\text{-}7)$$

the boundary conditions and the equation of continuity, Eq. (C) in Table 1.1, are satisfied. For a creeping flow under steady conditions, the left-hand side of the equation of motion is zero and Eqs. (D), (E), and (F) in Table 1.7 simplify to

$$r\text{-component} \quad 0 = \frac{\partial P}{\partial r}, \qquad (8.3\text{-}8)$$

$$\theta\text{-component} \quad 0 = \frac{\partial P}{\partial \theta}, \qquad (8.3\text{-}9)$$

$$\phi\text{-component} \quad 0 = \frac{\partial}{\partial r}\left(r^2 \frac{\partial v_\phi}{\partial r}\right) + \frac{1}{\sin\theta}\frac{\partial}{\partial\theta}\left(\sin\theta \frac{\partial v_\phi}{\partial\theta}\right) - \frac{v_\phi}{\sin^2\theta}. \quad (8.3\text{-}10)$$

Equations (8.3-8) and (8.3-9) indicate that pressure is constant throughout the fluid. Substitution of Eq. (8.3-7) into Eq. (8.3-10) yields

$$r^2 \frac{d^2 u}{dr^2} + 2r\frac{du}{dr} - 2u = 0. \qquad (8.3\text{-}11)$$

Mathematical Preliminary. An ordinary differential equation of the form

$$ax^2 \frac{d^2 y}{dx^2} + bx\frac{dy}{dx} + cy = 0,$$

where a, b, and c are constants, has the solution of the form

$$y = x^n.$$

Let us assume a solution of the form

$$u = r^n. \tag{8.3-12}$$

The first and second derivatives of u with respect to r are

$$\frac{du}{dr} = nr^{n-1} \quad \text{and} \quad \frac{d^2u}{dr^2} = n(n-1)r^{n-2}. \tag{8.3-13}$$

Substitution of Eqs. (8.3-12) and (8.3-13) into Eq. (8.3-11) leads to

$$n^2 + n - 2 = 0. \tag{8.3-14}$$

The solution of this quadratic equation gives

$$n_{1,2} = \frac{-1 \pm \sqrt{1+8}}{2} \quad \Rightarrow \quad n_1 = -2 \quad \text{and} \quad n_2 = 1. \tag{8.3-15}$$

Therefore, the solution is

$$u = \frac{C_1}{r^2} + C_2\, r, \tag{8.3-16}$$

or

$$v_\phi = \left(\frac{C_1}{r^2} + C_2\, r\right) \sin\theta. \tag{8.3-17}$$

Application of the boundary condition given by Eq. (8.3-6) indicates that $C_2 = 0$. The use of Eq. (8.3-5) gives $C_1 = \Omega R^3$. Therefore, the velocity distribution is expressed as

$$v_\phi = \frac{\Omega R^3 \sin\theta}{r^2}. \tag{8.3-18}$$

From Table 1.4,

$$\tau_{r\phi} = \tau_{\phi r} = -\mu r \frac{\partial}{\partial r}\left(\frac{v_\phi}{r}\right). \tag{8.3-19}$$

The use of Eq. (8.3-18) in Eq. (8.3-19) gives the shear stress distribution as

$$\tau_{r\phi} = 3\mu\Omega \sin\theta \left(\frac{R}{r}\right)^3. \tag{8.3-20}$$

The z-component of the torque is given by Eq. (4.2-29), i.e.,

$$T_z = \int_A \mathbf{e}_z \cdot [\mathbf{r} \times (\mathbf{n} \cdot \boldsymbol{\pi})] \, dA. \tag{8.3-21}$$

To calculate the torque applied by the rotating sphere on the fluid, the unit normal vector should be directed from the sphere to the fluid, i.e., $\mathbf{n} = \mathbf{e}_r$. From Eq. (4.2-28), the position vector in the spherical coordinate system is $\mathbf{r} = r\mathbf{e}_r$. Thus, the integrand in Eq. (8.3-21) is

$$
\begin{aligned}
\mathbf{e}_z \cdot [\mathbf{r} \times (\mathbf{n} \cdot \boldsymbol{\pi})] &= \mathbf{e}_z \cdot [r\mathbf{e}_r \times (P\mathbf{e}_r + \mathbf{e}_r \cdot \tau_{ij}\mathbf{e}_i\mathbf{e}_j] \\
&= \mathbf{e}_z \cdot [r\mathbf{e}_r \times (P\mathbf{e}_r + \tau_{rj}\mathbf{e}_j] \\
&= \mathbf{e}_z \cdot \epsilon_{rjk} r\tau_{rj}\mathbf{e}_k.
\end{aligned} \tag{8.3-22}
$$

From Eq. (A.6-17) in Appendix A, $\mathbf{e}_z = \cos\theta\,\mathbf{e}_r - \sin\theta\,\mathbf{e}_\theta$. Thus, Eq. (8.3-22) becomes

$$
\begin{aligned}
\mathbf{e}_z \cdot [\mathbf{r} \times (\mathbf{n} \cdot \boldsymbol{\pi})] &= (\cos\theta\,\mathbf{e}_r - \sin\theta\,\mathbf{e}_\theta) \cdot \epsilon_{rjk} r\tau_{rj}\mathbf{e}_k \\
&= \underbrace{\epsilon_{rjr} r\tau_{rj}\cos\theta}_{0} - \underbrace{\epsilon_{rj\theta} r\tau_{rj}\sin\theta}_{j=\phi \text{ and } \epsilon_{r\phi\theta}=-1} \\
&= r\tau_{r\phi}\sin\theta.
\end{aligned} \tag{8.3-23}
$$

The term A in Eq. (8.3-21) represents the surface of the sphere. From Eq. (A.6-14) in Appendix A, $dA = R^2 \sin\theta \, d\theta \, d\phi$. Therefore, Eq. (8.3-21) is expressed as

$$
\begin{aligned}
T_z &= \int_0^{2\pi} \int_0^{\pi} (r\tau_{r\phi}\sin\theta)|_{r=R} \, R^2 \sin\theta \, d\theta \, d\phi \\
&= 2\pi R^3 \int_0^{\pi} \tau_{r\phi}|_{r=R}\sin^2\theta \, d\theta.
\end{aligned} \tag{8.3-24}
$$

Substitution of Eq. (8.3-20) into Eq. (8.3-24) and integration lead to

$$T_z = 8\pi\mu\Omega R^3, \tag{8.3-25}$$

which represents the torque applied by the rotating sphere on the fluid. On the other hand, the torque exerted by the fluid on the sphere is $-8\pi\mu\Omega R^3$.

Comment: The creeping flow assumption implies that the inertial terms in the equation of motion are considered zero. However, if the postulate expressed by Eq. (8.3-7) is correct, the inertial terms in the ϕ-component of the equation of motion vanish and we still obtain Eq. (8.3-10) without making the creeping flow assumption. Therefore, what is the purpose of making such an assumption?

Without the creeping flow assumption, the r- and θ-components of the equation of motion simplify to

$$r\text{-component} \qquad \rho\frac{v_\phi^2}{r} = \frac{\partial P}{\partial r}, \qquad (8.3\text{-}26)$$

$$\theta\text{-component} \qquad \rho v_\phi^2 \cot\theta = \frac{\partial P}{\partial \theta}. \qquad (8.3\text{-}27)$$

Substitution of Eq. (8.3-18) into Eq. (8.3-26) and integration give

$$P = -\frac{\rho\Omega^2 R^6 \sin^2\theta}{r^4} + K(\theta). \qquad (8.3\text{-}28)$$

Substitution of Eqs. (8.3-18) and (8.3-28) into Eq. (8.3-27) and rearrangement yield

$$\underbrace{\frac{dK}{d\theta}}_{\text{function of }\theta} = \underbrace{\frac{3\rho\Omega^2 R^6 \sin\theta\cos\theta}{r^4}}_{\text{function of }r\text{ and }\theta}, \qquad (8.3\text{-}29)$$

which is physically impossible.

8.3.1 *Alternative solution*

Let us postulate the velocity component in the ϕ-direction as

$$v_\phi = rf(r)\sin\theta, \qquad (8.3\text{-}30)$$

and resolve Eq. (8.3-10) for the velocity distribution. Substitution of Eq. (8.3-30) into Eq. (8.3-10) and rearrangement give

$$\frac{d^2 f}{dr^2} + \frac{4}{r}\frac{df}{dr} = 0. \qquad (8.3\text{-}31)$$

The integrating factor, I, is

$$I = \exp\left(4\int\frac{dr}{r}\right) = r^4. \qquad (8.3\text{-}32)$$

Multiplication of Eq. (8.3-31) by the integrating factor leads to

$$r^4\frac{d^2 f}{dr^2} + 4r^3\frac{df}{dr} = 0 \quad\Rightarrow\quad \frac{d}{dr}\left(r^4\frac{df}{dr}\right) = 0. \qquad (8.3\text{-}33)$$

Therefore,

$$r^4 \frac{df}{dr} = K_1 \quad \Rightarrow \quad \frac{df}{dr} = \frac{K_1}{r^4}. \tag{8.3-34}$$

Integration yields

$$f = -\frac{K_1}{3r^3} + K_2 \Rightarrow v_\phi = \left(-\frac{K_1}{3r^2} + K_2 r \right) \sin\theta. \tag{8.3-35}$$

If we let $K_1' = -K_1/3$, Eq. (8.3-5) transforms into

$$v_\phi = \left(\frac{K_1'}{r^2} + K_2 r \right) \sin\theta, \tag{8.3-36}$$

which is identical to Eq. (8.3-17).

8.4   Flow Between Rotating Concentric Spheres

A Newtonian fluid is contained between two concentric spheres as shown in Figure 8.3. While the inner sphere is held stationary, the outer sphere rotates at a constant angular velocity Ω.

The procedure presented in Section 8.3 can be used to determine the velocity distribution. Thus, from Eq. (8.3-17),

$$\frac{v_\phi}{\sin\theta} = \frac{C_1}{r^2} + C_2\, r. \tag{8.4-1}$$

The boundary conditions are

$$\text{at } r = \kappa R \quad v_\phi = 0, \tag{8.4-2}$$

$$\text{at } r = R \quad v_\phi = \Omega R \sin\theta. \tag{8.4-3}$$

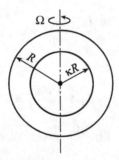

Figure 8.3   Flow between concentric spheres.

Application of the boundary conditions yields the following equations:

$$0 = \frac{C_1}{\kappa^2 R^2} + C_2 \kappa R, \tag{8.4-4}$$

$$\Omega R = \frac{C_1}{R^2} + C_2 R. \tag{8.4-5}$$

In matrix notation, Eqs. (8.4-4) and (8.4-5) are expressed in the form of

$$\begin{pmatrix} 0 \\ \Omega R \end{pmatrix} = \begin{pmatrix} 1/\kappa^2 R^2 & \kappa R \\ 1/R^2 & R \end{pmatrix} \begin{pmatrix} C_1 \\ C_2 \end{pmatrix}. \tag{8.4-6}$$

The constants are evaluated with the help of Cramer's rule as follows:

$$C_1 = \frac{\begin{vmatrix} 0 & \kappa R \\ \Omega R & R \end{vmatrix}}{\begin{vmatrix} 1/\kappa^2 R^2 & \kappa R \\ 1/R^2 & R \end{vmatrix}} = -\frac{\Omega \kappa R^2}{1/\kappa^2 R - \kappa/R} = -\frac{\Omega \kappa^3 R^3}{1 - \kappa^3}, \tag{8.4-7}$$

$$C_2 = \frac{\begin{vmatrix} 1/\kappa^2 R^2 & 0 \\ 1/R^2 & \Omega R \end{vmatrix}}{\begin{vmatrix} 1/\kappa^2 R^2 & \kappa R \\ 1/R^2 & R \end{vmatrix}} = \frac{\Omega/\kappa^2 R}{1/\kappa^2 R - \kappa/R} = \frac{\Omega}{1 - \kappa^3}. \tag{8.4-8}$$

Substitution of Eqs. (8.4-7) and (8.4-8) into Eq. (8.4-1) and rearrangement result in

$$\frac{v_\phi}{r \sin \theta} = \frac{\Omega}{1 - \kappa^3} \left[1 - \left(\frac{\kappa R}{r} \right)^3 \right]. \tag{8.4-9}$$

From Table 1.4,

$$\tau_{r\phi} = \tau_{\phi r} = -\mu r \frac{\partial}{\partial r} \left(\frac{v_\phi}{r} \right). \tag{8.4-10}$$

The use of Eq. (8.4-9) in Eq. (8.4-10) gives the shear stress distribution as

$$\tau_{r\phi} = -\frac{3\mu\Omega \sin \theta}{1 - \kappa^3} \left(\frac{\kappa R}{r} \right)^3. \tag{8.4-11}$$

The torque exerted by the fluid on the surface of the inner sphere is given by

$$T_z = -\int_0^{2\pi} \int_0^{\pi} (r\tau_{r\phi} \sin \theta)|_{r=\kappa R} (\kappa R)^2 \sin \theta \, d\theta \, d\phi$$

$$= -2\pi(\kappa R)^3 \int_0^{\pi} \tau_{r\phi}|_{r=\kappa R} \sin^2 \theta \, d\theta. \tag{8.4-12}$$

Substitution of Eq. (8.4-11) into Eq. (8.4-12) and integration lead to

$$T_z = \frac{8\pi\mu\Omega(\kappa R)^3}{1 - \kappa^3}. \tag{8.4-13}$$

8.5   Flow of a Newtonian Fluid Between Rotating Cones

Consider steady, creeping flow of an incompressible Newtonian fluid contained in the gap between two coaxial cones as shown in Figure 8.4. The angular velocities of the inner and outer cones are Ω_i and Ω_o, respectively.

Rotational symmetry around the z-axis implies that the components of velocity and pressure are independent of ϕ. The boundary conditions are

$$\text{at } \theta = \theta_i \qquad v_r = v_\theta = 0 \quad \text{and} \quad v_\phi = \Omega_i r \sin \theta_i, \qquad (8.5\text{-}1)$$

$$\text{at } \theta = \theta_o \qquad v_r = v_\theta = 0 \quad \text{and} \quad v_\phi = \Omega_o r \sin \theta_o. \qquad (8.5\text{-}2)$$

Based on the boundary conditions one can postulate the velocity components in two forms.

Case (i): Let us postulate

$$v_r = v_\theta = 0 \qquad \text{and} \qquad v_\phi = r\,u(\theta). \qquad (8.5\text{-}3)$$

The components of the equation of motion, Eqs. (D), (E), and (F) in Table 1.7, simplify to

$$r\text{-component} \quad 0 = \frac{\partial P}{\partial r}, \qquad (8.5\text{-}4)$$

$$\theta\text{-component} \quad 0 = \frac{\partial P}{\partial \theta}, \qquad (8.5\text{-}5)$$

$$\phi\text{-component} \quad 0 = \frac{\partial}{\partial r}\left(r^2 \frac{\partial v_\phi}{\partial r}\right) + \frac{1}{\sin \theta} \frac{\partial}{\partial \theta}\left(\sin \theta \frac{\partial v_\phi}{\partial \theta}\right) - \frac{v_\phi}{\sin^2 \theta}. \quad (8.5\text{-}6)$$

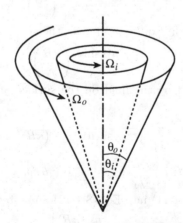

Figure 8.4   Flow between rotating cones.

Substitution of Eq. (8.5-3) into Eq. (8.5-6) yields

$$\frac{d^2u}{d\theta^2} + \frac{\cos\theta}{\sin\theta}\frac{du}{d\theta} + \left(2 - \frac{1}{\sin^2\theta}\right)u = 0. \tag{8.5-7}$$

Case (ii): Let us postulate

$$v_r = v_\theta = 0 \quad \text{and} \quad v_\phi = r\,f(\theta)\sin\theta. \tag{8.5-8}$$

Substitution of Eq. (8.5-8) into Eq. (8.5-6) results in

$$\frac{d^2f}{d\theta^2} + \frac{3\cos\theta}{\sin\theta}\frac{df}{d\theta} = 0. \tag{8.5-9}$$

Examination of Eqs. (8.5-7) and (8.5-9) indicates that the latter is simpler to solve. Thus, the integrating factor, I, is

$$I = \exp\left(3\int\frac{\cos\theta}{\sin\theta}\,d\theta\right) = \sin^3\theta. \tag{8.5-10}$$

Multiplication of Eq. (8.5-9) by the integrating factor leads to

$$\sin^3\theta\,\frac{d^2f}{d\theta^2} + 3\sin^2\theta\cos\theta\,\frac{df}{d\theta} = 0 \quad\Rightarrow\quad \frac{d}{d\theta}\left(\sin^3\theta\,\frac{df}{d\theta}\right) = 0. \tag{8.5-11}$$

Therefore,

$$\sin^3\theta\,\frac{df}{d\theta} = C_1 \quad\Rightarrow\quad \frac{df}{d\theta} = \frac{C_1}{\sin^3\theta}. \tag{8.5-12}$$

Integration yields

$$f = -\frac{C_1}{2}\frac{\cos\theta}{\sin^2\theta} + \frac{C_1}{2}\ln\left[\tan(\theta/2)\right] + C_2, \tag{8.5-13}$$

or

$$f = -K_1\frac{\cos\theta}{\sin^2\theta} + K_1\ln\left[\tan(\theta/2)\right] + C_2, \tag{8.5-14}$$

where $K_1 = C_1/2$. The solution becomes

$$\frac{v_\phi}{r\sin\theta} = -K_1\frac{\cos\theta}{\sin^2\theta} + K_1\ln\left[\tan(\theta/2)\right] + C_2. \tag{8.5-15}$$

Application of Eq. (8.5-1) gives

$$C_2 = \Omega_i + K_1\left\{\frac{\cos\theta_i}{\sin^2\theta_i} - \ln\left[\tan(\theta_i/2)\right]\right\}. \tag{8.5-16}$$

Substitution of Eq. (8.5-16) into Eq. (8.5-15) and rearrangement yield

$$\frac{v_\phi}{r\sin\theta} = \Omega_i + K_1 \left\{ \frac{\cos\theta_i}{\sin^2\theta_i} - \frac{\cos\theta}{\sin^2\theta} + \ln\left[\frac{\tan(\theta/2)}{\tan(\theta_i/2)}\right] \right\}. \tag{8.5-17}$$

Application of Eq. (8.5-2) gives K_1 as

$$K_1 = (\Omega_o - \Omega_i) \left\{ \frac{\cos\theta_i}{\sin^2\theta_i} - \frac{\cos\theta_o}{\sin^2\theta_o} + \ln\left[\frac{\tan(\theta_o/2)}{\tan(\theta_i/2)}\right] \right\}^{-1}. \tag{8.5-18}$$

Therefore, the velocity distribution is

$$\frac{\dfrac{v_\phi}{r\sin\theta} - \Omega_i}{\Omega_o - \Omega_i} = \frac{\dfrac{\cos\theta_i}{\sin^2\theta_i} - \dfrac{\cos\theta}{\sin^2\theta} + \ln\left[\dfrac{\tan(\theta/2)}{\tan(\theta_i/2)}\right]}{\dfrac{\cos\theta_i}{\sin^2\theta_i} - \dfrac{\cos\theta_o}{\sin^2\theta_o} + \ln\left[\dfrac{\tan(\theta_o/2)}{\tan(\theta_i/2)}\right]}. \tag{8.5-19}$$

8.6   Flow of a Power-Law Fluid Between Rotating Cones

Let us repeat the analysis presented in Section 8.5 for a power-law fluid. Postulating the velocity components as

$$v_r = v_\theta = 0 \qquad \text{and} \qquad v_\phi = r\, f(\theta)\sin\theta, \tag{8.6-1}$$

Table 1.4 indicates that the only non-zero shear stress component is

$$\tau_{\theta\phi} = \tau_{\phi\theta} = -\eta\, \frac{\sin\theta}{r}\, \frac{\partial}{\partial\theta}\left(\frac{v_\phi}{\sin\theta}\right). \tag{8.6-2}$$

The use of Eq. (8.6-1) in Eq. (8.6-2) gives

$$\tau_{\theta\phi} = -\eta\sin\theta\, \frac{df}{d\theta}. \tag{8.6-3}$$

From Eq. (C) in Table 1.7, the ϕ-component of the equation of motion simplifies to

$$\frac{d\tau_{\theta\phi}}{d\theta} + 2\cot\theta\,\tau_{\theta\phi} = 0. \tag{8.6-4}$$

Equation (8.6-4) is a separable equation and can be expressed as

$$\frac{d\tau_{\theta\phi}}{\tau_{\theta\phi}} = -2\cot\theta\, d\theta. \tag{8.6-5}$$

Integration of Eq. (8.6-5) gives

$$\ln\tau_{\theta\phi} = -2\ln(\sin\theta) + \ln C_1, \tag{8.6-6}$$

or

$$\tau_{\theta\phi} \sin^2 \theta = C_1. \tag{8.6-7}$$

For a power-law fluid, the non-Newtonian viscosity is given by Eq. (7.1-6). From Eq. (C) in Table 7.1,

$$\frac{1}{2} (\dot{\gamma} : \dot{\gamma}) = \left[\frac{\sin \theta}{r} \frac{\partial}{\partial \theta} \left(\frac{v_\phi}{\sin \theta} \right) \right]^2 = \left(\sin \theta \frac{df}{d\theta} \right)^2. \tag{8.6-8}$$

Hence,

$$\dot{\gamma} = \left| \sqrt{\frac{1}{2} (\dot{\gamma} : \dot{\gamma})} \right| = \left| \sin \theta \frac{df}{d\theta} \right|. \tag{8.6-9}$$

Substitution of Eq. (8.6-9) into Eq. (7.1-6) yields

$$\eta = m \left| \sin \theta \frac{df}{d\theta} \right|^{n-1}. \tag{8.6-10}$$

The use of Eq. (8.6-10) in Eq. (8.6-3) leads to the following expression for the shear stress:

$$\tau_{\theta\phi} = - m \left| \sin \theta \frac{df}{d\theta} \right|^{n-1} \sin \theta \frac{df}{d\theta}. \tag{8.6-11}$$

Case (i) $\Omega_o > \Omega_i$:
In this case, v_θ increases with an increase in θ. Thus, $df/d\theta > 0$ and Eq. (8.6-11) becomes

$$\tau_{\theta\phi} = - m \left(\sin \theta \frac{df}{d\theta} \right)^n. \tag{8.6-12}$$

Substitution of Eq. (8.6-12) into Eq. (8.6-7) and rearrangement give

$$\frac{df}{d\theta} = K_1 \csc^{2s+1} \theta, \tag{8.6-13}$$

where

$$s = \frac{1}{n} \quad \text{and} \quad K_1 = \left(-\frac{C_1}{m} \right)^s. \tag{8.6-14}$$

Integration of Eq. (8.6-13) leads to

$$f = K_1 \int_0^\theta \csc^{2s+1} \theta \, d\theta + C_2. \tag{8.6-15}$$

Application of the boundary conditions

$$\text{at } \theta = \theta_i \quad f = \Omega_i, \tag{8.6-16}$$

$$\text{at } \theta = \theta_o \quad f = \Omega_o, \tag{8.6-17}$$

results in the following distribution for f:

$$\frac{f - \Omega_i}{\Omega_o - \Omega_i} = \frac{\int_{\theta_i}^{\theta} \csc^{2s+1}\theta \, d\theta}{\int_{\theta_i}^{\theta_o} \csc^{2s+1}\theta \, d\theta}. \tag{8.6-18}$$

Case (ii) $\Omega_o < \Omega_i$:
In this case, v_θ decreases with an increase in θ. Thus, $df/d\theta < 0$ and Eq. (8.6-11) becomes

$$\tau_{\theta\phi} = m\left(-\sin\theta\,\frac{df}{d\theta}\right)^n. \tag{8.6-19}$$

Substitution of Eq. (8.6-19) into Eq. (8.6-7) and rearrangement give

$$\frac{df}{d\theta} = K_2 \csc^{2s+1}\theta, \tag{8.6-20}$$

where

$$K_2 = -\left(\frac{C_1}{m}\right)^s. \tag{8.6-21}$$

Integration of Eq. (8.6-20) leads to

$$f = K_2 \int_0^\theta \csc^{2s+1}\theta \, d\theta + C_3. \tag{8.6-22}$$

Application of the boundary conditions given by Eqs. (8.6-16) and (8.6-17) leads to

$$\frac{\Omega_i - f}{\Omega_i - \Omega_o} = \frac{\int_{\theta_i}^{\theta} \csc^{2s+1}\theta \, d\theta}{\int_{\theta_i}^{\theta_o} \csc^{2s+1}\theta \, d\theta}. \tag{8.6-23}$$

Therefore, velocity decreases from Ω_i to Ω_o.

Case (iii) $\Omega_o = \Omega_i$:
In this case, both Eqs. (8.6-18) and (8.6-23) imply that

$$f = \Omega_i. \tag{8.6-24}$$

The use of Eq. (8.6-24) in Eq. (8.6-3) gives

$$\tau_{\theta\phi} = 0, \tag{8.6-25}$$

implying that a power-law fluid rotates around the z-axis without deformation as if it is a solid body.

- **Simplification for a Newtonian fluid**

  When $n = s = 1$, Eqs. (8.6-18) and (8.6-23) simplify to

$$\frac{f - \Omega_i}{\Omega_o - \Omega_i} = \frac{\int_{\theta_i}^{\theta} \csc^3 \theta \, d\theta}{\int_{\theta_i}^{\theta_o} \csc^3 \theta \, d\theta}. \tag{8.6-26}$$

  Noting that

$$\int \csc^3 \theta \, d\theta = \frac{1}{2} \left\{ \ln \left[\tan \left(\frac{\theta}{2} \right) \right] - \frac{\cos \theta}{\sin^2 \theta} \right\}, \tag{8.6-27}$$

  Eq. (8.6-26) reduces to Eq. (8.5-19).

8.7 Flow in a Tapered Tube

Consider steady, creeping flow of an incompressible Newtonian fluid in a tapered tube (or circular cone) as shown in Figure 8.5. The vertex of the cone coincides with the origin O, and the axis of the cone is oriented along the z-axis. The surface of the cone is taken as the surface of revolution $\theta = \text{constant}$. The streamlines are straight lines radiating from the origin and therefore given by the values $\theta = \text{constant}$.

Postulating

$$v_r = v_r(r, \theta) \qquad \text{and} \qquad v_\theta = v_\phi = 0, \tag{8.7-1}$$

the equation of continuity, Eq. (C) in Table 1.1, becomes

$$\frac{\partial}{\partial r}(r^2 v_r) = 0 \quad \Rightarrow \quad r^2 v_r = u(\theta) \quad \Rightarrow \quad v_r = \frac{u(\theta)}{r^2}. \tag{8.7-2}$$

On the other hand, the r- and θ-components of the equation of motion, Eqs. (D) and (E) in Table 1.7, simplify to

$$0 = -\frac{\partial P}{\partial r} + \mu \left[\frac{1}{r^2} \frac{\partial}{\partial r} \left(r^2 \frac{\partial v_r}{\partial r} \right) + \frac{1}{r^2 \sin \theta} \frac{\partial}{\partial \theta} \left(\sin \theta \frac{\partial v_r}{\partial \theta} \right) - \frac{2}{r^2} v_r \right], \tag{8.7-3}$$

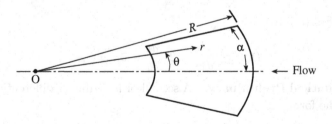

Figure 8.5   Flow in a tapered tube.

$$0 = -\frac{1}{r}\frac{\partial P}{\partial \theta} + \frac{2\mu}{r^2}\frac{\partial v_r}{\partial \theta}. \tag{8.7-4}$$

Substitution of Eq. (8.7-2) into Eqs. (8.7-3) and (8.7-4) leads to the following equations:

$$\frac{\partial P}{\partial r} = \frac{\mu}{r^4 \sin\theta}\frac{d}{d\theta}\left(\sin\theta\frac{du}{d\theta}\right), \tag{8.7-5}$$

$$\frac{\partial P}{\partial \theta} = \frac{2\mu}{r^3}\frac{du}{d\theta}. \tag{8.7-6}$$

Differentiation of Eq. (8.7-5) with respect to θ and differentiation of Eq. (8.7-6) with respect to r yield

$$\frac{\partial^2 P}{\partial \theta\,\partial r} = \frac{\mu}{r^4}\frac{d}{d\theta}\left[\frac{1}{\sin\theta}\frac{d}{d\theta}\left(\sin\theta\frac{du}{d\theta}\right)\right], \tag{8.7-7}$$

$$\frac{\partial^2 P}{\partial r\,\partial \theta} = 2\mu\frac{du}{d\theta}\frac{d}{dr}\left(\frac{1}{r^3}\right) = -\frac{6\mu}{r^4}\frac{du}{d\theta}. \tag{8.7-8}$$

Subtraction of Eq. (8.7-8) from Eq. (8.7-7) eliminates the pressure terms. The resulting equation is

$$\frac{d}{d\theta}\left[\frac{1}{\sin\theta}\frac{d}{d\theta}\left(\sin\theta\frac{du}{d\theta}\right)\right] + 6\frac{du}{d\theta} = 0. \tag{8.7-9}$$

Integration of Eq. (8.7-9) with respect to θ results in

$$\frac{1}{\sin\theta}\frac{d}{d\theta}\left(\sin\theta\frac{du}{d\theta}\right) + 6u = C_1, \tag{8.7-10}$$

where C_1 is an integration constant.

The transformation

$$x = \cos\theta, \tag{8.7-11}$$

reduces Eq. (8.7-10) to

$$(1-x^2)\frac{d^2u}{dx^2} - 2x\frac{du}{dx} + 6u = C_1. \tag{8.7-12}$$

Mathematical Preliminary. A second-order ordinary differential equation of the form

$$(1-x^2)\frac{d^2 f}{dx^2} - 2x\frac{df}{dx} + n(n+1)f = 0,$$

where n is a real constant, is known as Legendre's equation of order n. When n is an integer, the general solution is given by

$$f = C_1 P_n(x) + C_2 Q_n(x),$$

where $P_n(x)$ and $Q_n(x)$ are the Legendre functions of the first and second kinds, respectively.

Noting that

$$n(n+1) = 6 \quad \Rightarrow \quad n = 2, \tag{8.7-13}$$

the homogeneous equation

$$(1 - x^2)\frac{d^2 u}{dx^2} - 2x\frac{du}{dx} + 6u = 0 \tag{8.7-14}$$

has the following solution:

$$u = A\,P_2(x) + B\,Q_2(x). \tag{8.7-15}$$

In Eq. (8.7-15), A and B are constants and the Legendre functions are given as

$$P_2(x) = \frac{3x^2 - 1}{2} \quad \text{and} \quad Q_2(x) = \left(\frac{3x^2 - 1}{2}\right)\ln\left(\frac{1+x}{1-x}\right) - \frac{3x}{2}. \tag{8.7-16}$$

The particular solution of Eq. (8.7-12) is

$$u_p = \frac{C_1}{6}. \tag{8.7-17}$$

The summation of homogeneous and particular solutions gives the general solution as

$$u(\theta) = A\,P_2(\cos\theta) + B\,Q_2(\cos\theta) + \frac{C_1}{6}. \tag{8.7-18}$$

At the axis of the cone, i.e., $\theta = 0$, v_r is finite. According to Eq. (8.7-2), $u(\theta)$ must also be finite at $\theta = 0$. Noting that

$$\theta = 0 \quad \Rightarrow \quad x = \cos\theta = 1 \quad \Rightarrow \quad Q_2(x) = \infty, \tag{8.7-19}$$

B must be zero and the solution becomes

$$u(\theta) = A\,P_2(\cos\theta) + \frac{C_1}{6}. \tag{8.7-20}$$

Since the velocity of the liquid vanishes at the wall, application of the no-slip boundary condition, i.e.,

$$\text{at} \quad \theta = \alpha \quad u(\theta) = 0, \tag{8.7-21}$$

gives the solution as

$$u(\theta) = \frac{3A}{2} \left(\cos^2 \theta - \cos^2 \alpha \right). \tag{8.7-22}$$

The use of the trigonometric identity

$$\cos^2 \theta = \frac{1 + \cos 2\theta}{2}, \tag{8.7-23}$$

transforms Eq. (8.7-22) to

$$u(\theta) = \frac{3A}{4} \left(\cos 2\theta - \cos 2\alpha \right). \tag{8.7-24}$$

Substitution of Eq. (8.7-24) into Eq. (8.7-2) gives the velocity distribution as

$$v_r = \frac{3A}{4r^2} \left(\cos 2\theta - \cos 2\alpha \right). \tag{8.7-25}$$

Note that the constant A will be determined later.

Substitution of Eq. (8.7-24) into Eq. (8.7-6) results in

$$\frac{\partial P}{\partial \theta} = -\frac{3A\mu}{r^3} \sin 2\theta. \tag{8.7-26}$$

Integration of Eq. (8.7-26) yields

$$P = \frac{3A\mu}{2r^3} \cos 2\theta + K(r). \tag{8.7-27}$$

Substitution of Eqs. (8.7-24) and (8.7-27) into Eq. (8.7-5) leads to

$$\frac{dK}{dr} = -\frac{3A\mu}{2r^4}. \tag{8.7-28}$$

Integration of Eq. (8.7-28) gives

$$K = \frac{A\mu}{2r^3} + C_2. \tag{8.7-29}$$

Substitution of Eq. (8.7-29) into Eq. (8.7-27) gives the pressure distribution as

$$P = \frac{3}{2} \frac{A\mu}{r^3} \left(\cos 2\theta + 1/3 \right) + C_2. \tag{8.7-30}$$

Application of the boundary condition

$$\text{at } r = R \quad P = P_{in},\tag{8.7-31}$$

gives the pressure distribution as

$$P - P_{in} = \frac{3}{2}A\mu \left(\frac{\cos 2\theta + 1/3}{r^3} - \frac{\cos 2\theta + 1/3}{R^3} \right).\tag{8.7-32}$$

The volumetric flow rate is

$$Q = \int_0^{2\pi} \int_0^{\alpha} v_r r^2 \sin\theta \, d\theta \, d\phi.\tag{8.7-33}$$

Substitution of Eq. (8.7-25) into Eq. (8.7-33) and integration yield

$$Q = A\pi(1 - \cos\alpha)^2(1 + 2\cos\alpha).\tag{8.7-34}$$

Therefore, the constant A is expressed as

$$A = \frac{Q}{\pi(1 - \cos\alpha)^2(1 + 2\cos\alpha)}.\tag{8.7-35}$$

Substitution of Eq. (8.7-35) into Eqs. (8.7-25) and (8.7-32) gives the velocity and pressure distributions as

$$v_r = \frac{3Q}{4\pi r^2} \frac{(\cos 2\theta - \cos 2\alpha)}{(1 - \cos\alpha)^2(1 + 2\cos\alpha)}$$

$$= -\frac{3Q}{2\pi r^2} \frac{\sin^2\alpha}{(1 - \cos\alpha)^2(1 + 2\cos\alpha)} \left[1 - \left(\frac{\sin\theta}{\sin\alpha} \right)^2 \right],\tag{8.7-36}$$

$$P - P_{in} = \frac{3\mu Q}{2\pi} \frac{1}{(1 - \cos\alpha)^2(1 + 2\cos\alpha)} \left(\frac{\cos 2\theta + 1/3}{r^3} - \frac{\cos 2\theta + 1/3}{R^3} \right).\tag{8.7-37}$$

Chapter 9

Lubrication Approximation

As stated in Section 1.1.2, a *path line* is the curve or path traced by a particle during flow. A *streamline* is the curve whose tangent at any point is in the direction of the velocity at that point. In the case of steady-state flow, path lines and streamlines are identical.

For steady flow in conduits with constant cross-sectional area, such as parallel plates and circular tubes, path lines (or streamlines) are parallel to the conduit surfaces and the tangent drawn to path lines has only one component, i.e., one-dimensional flow. For flow in conduits with varying cross-sections, however, the conduit surfaces are not parallel to each other. As a result, path lines are curved and the tangent drawn to path lines has two components, resulting in two-dimensional flow. When the surfaces are "nearly parallel" to each other, this complicated problem can be simplified by making use of the *lubrication approximation*.[1]

9.1   Flow of a Newtonian Fluid in a Tapered Tube

For flow of an incompressible Newtonian fluid under steady conditions through a circular tube of radius R as shown in Figure 5.3(a), the velocity distribution and the volumetric flow rate are given as

$$v_z = \frac{(\mathcal{P}_o - \mathcal{P}_L)R^2}{4\mu L}\left[1 - \left(\frac{r}{R}\right)^2\right], \tag{9.1-1}$$

$$Q = \frac{\pi(\mathcal{P}_o - \mathcal{P}_L)R^4}{8\mu L}. \tag{9.1-2}$$

---

[1]This approximation was first used by Reynolds (1886) in a study of lubrication.

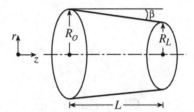

Figure 9.1   Flow through a tapered tube.

Now consider the flow in a tapered tube under steady conditions as shown in Figure 9.1. Since the tube surfaces are not parallel to each other, two velocity components, v_r and v_z, should be taken into consideration.

The following analysis is valid when $R_o/L \ll 1$ and the taper angle β is small, i.e., $(R_o - R_L)/L \ll 1$. Postulating $v_r = v_r(r, z)$, $v_\theta = 0$, $v_z = v_z(r, z)$, and $P = P(r, z)$, the equation of continuity, Eq. (B) in Table 1.1, and the r- and z-components of the equation of motion, Eqs. (D) and (F) in Table 1.6, take the form

$$\frac{1}{r}\frac{\partial}{\partial r}(rv_r) + \frac{\partial v_z}{\partial z} = 0, \tag{9.1-3}$$

$$\rho\left(v_r\frac{\partial v_r}{\partial r} + v_z\frac{\partial v_r}{\partial z}\right) = -\frac{\partial P}{\partial r} + \mu\left\{\frac{\partial}{\partial r}\left[\frac{1}{r}\frac{\partial}{\partial r}(rv_r)\right] + \frac{\partial^2 v_r}{\partial z^2}\right\}, \tag{9.1-4}$$

$$\rho\left(v_r\frac{\partial v_z}{\partial r} + v_z\frac{\partial v_z}{\partial z}\right) = -\frac{\partial P}{\partial z} + \mu\left[\frac{1}{r}\frac{\partial}{\partial r}\left(r\frac{\partial v_z}{\partial r}\right) + \frac{\partial^2 v_z}{\partial z^2}\right]. \tag{9.1-5}$$

The modified pressure P in Eqs. (9.1-4) and (9.1-5) is defined by Eq. (5.2-25) with $\cos\alpha = 1$.

Since R_o and R_L do not differ significantly, let R_o be the characteristic length in the r-direction over which v_z changes from zero to some maximum value. The characteristic length in the z-direction, on the other hand, is the length of the tube L, where $R_o \ll L$. Therefore, we can write

$$O(r) \sim R_o \quad \text{and} \quad O(z) \sim L. \tag{9.1-6}$$

The characteristic velocity in the z-direction is the average velocity $\langle v_z \rangle$, which is the volumetric flow rate divided by the cross-sectional area, and we can propose

$$O(v_z) \sim \langle v_z \rangle. \tag{9.1-7}$$

Let U be the characteristic velocity in the r-direction. The order of magnitude of U can be estimated by using the equation of continuity, Eq. (9.1-3). The velocity gradients in the equation of continuity have the following order of magnitude:

$$O\left[\frac{1}{r}\frac{\partial}{\partial r}(rv_r)\right] \sim \frac{U}{R_o}, \tag{9.1-8}$$

$$O\left(\frac{\partial v_z}{\partial z}\right) \sim \frac{\langle v_z\rangle}{L}. \tag{9.1-9}$$

If these two terms are retained in the equation of continuity, they must have the same order of magnitude, i.e.,

$$\frac{U}{R_o} \sim \frac{\langle v_z\rangle}{L} \quad \Rightarrow \quad O(U) \sim \langle v_z\rangle\frac{R_o}{L}. \tag{9.1-10}$$

Since $R_o \ll L$, Eq. (9.1-10) implies that $U \ll \langle v_z\rangle$.

In the r-component of the equation of motion, Eq. (9.1-4), the orders of magnitude of the viscous terms are

$$O\left\{\mu\frac{\partial}{\partial r}\left[\frac{1}{r}\frac{\partial}{\partial r}(rv_r)\right]\right\} \sim \frac{\mu\langle v_z\rangle}{R_o L} \quad \text{and} \quad O\left(\mu\frac{\partial^2 v_r}{\partial z^2}\right) \sim \frac{\mu\langle v_z\rangle R_o}{L^3}. \tag{9.1-11}$$

or

$$\frac{O\left\{\mu\dfrac{\partial}{\partial r}\left[\dfrac{1}{r}\dfrac{\partial}{\partial r}(rv_r)\right]\right\}}{O\left(\mu\dfrac{\partial^2 v_r}{\partial z^2}\right)} \sim \left(\frac{L}{R_o}\right)^2 \gg 1. \tag{9.1-12}$$

Equation (9.1-12) implies that the term $\mu(\partial^2 v_r/\partial z^2)$ is very small and can be neglected.

Similarly, in the z-component of the equation of motion, Eq. (9.1-5), the order of magnitude analysis leads to

$$O\left[\frac{\mu}{r}\frac{\partial}{\partial r}\left(r\frac{\partial v_z}{\partial r}\right)\right] \sim \frac{\mu\langle v_z\rangle}{R_o^2} \quad \text{and} \quad O\left(\mu\frac{\partial^2 v_z}{\partial z^2}\right) \sim \frac{\mu\langle v_z\rangle}{L^2}, \tag{9.1-13}$$

or

$$\frac{\dfrac{\mu}{r}\dfrac{\partial}{\partial r}\left(r\dfrac{\partial v_z}{\partial r}\right)}{\mu\dfrac{\partial^2 v_z}{\partial z^2}} \sim \left(\frac{L}{R_o}\right)^2 \gg 1. \tag{9.1-14}$$

Therefore, the term $\mu(\partial^2 v_z/\partial z^2)$ is very small and can also be neglected.

With these simplifications, Eqs. (9.1-4) and (9.1-5) reduce to

$$\underbrace{v_r \frac{\partial v_r}{\partial r} + v_z \frac{\partial v_r}{\partial z}}_{O(\langle v_z \rangle^2 R_o/L^2)} = -\frac{1}{\rho} \frac{\partial P}{\partial r} + \underbrace{\nu \frac{\partial}{\partial r} \left[\frac{1}{r} \frac{\partial}{\partial r}(r v_r) \right]}_{O(\nu \langle v_z \rangle / R_o L)}, \tag{9.1-15}$$

$$\underbrace{v_r \frac{\partial v_z}{\partial r} + v_z \frac{\partial v_z}{\partial z}}_{O(\langle v_z \rangle^2/L)} = -\frac{1}{\rho} \frac{\partial P}{\partial z} + \underbrace{\frac{\nu}{r} \frac{\partial}{\partial r} \left(r \frac{\partial v_z}{\partial r} \right)}_{O(\nu \langle v_z \rangle / R_o^2)}. \tag{9.1-16}$$

The term ν in Eqs. (9.1-15) and (9.1-16) is the kinematic viscosity, defined by μ/ρ. Equations (9.1-15) and (9.1-16) consist of inertial, pressure, and viscous terms and they must be of comparable magnitude. Consequently, either the orders of magnitude of the pressure and inertial terms or the pressure and viscous terms must be equal to each other. The order of magnitude of the pressure term in Eq. (9.1-15) is

$$O\left(\frac{1}{\rho} \frac{\partial P}{\partial r} \right) \sim \frac{\langle v_z \rangle^2 R_o}{L^2} \quad \text{or} \quad O\left(\frac{1}{\rho} \frac{\partial P}{\partial r} \right) \sim \frac{\nu \langle v_z \rangle}{R_o L}. \tag{9.1-17}$$

On the other hand, the order of magnitude of the pressure term in Eq. (9.1-16) is

$$O\left(\frac{1}{\rho} \frac{\partial P}{\partial z} \right) \sim \frac{\langle v_z \rangle^2}{L} \quad \text{or} \quad O\left(\frac{1}{\rho} \frac{\partial P}{\partial z} \right) \sim \frac{\nu \langle v_z \rangle}{R_o^2}. \tag{9.1-18}$$

Therefore,

$$\frac{\partial P/\partial r}{\partial P/\partial z} \sim \frac{\rho \langle v_z \rangle^2 R_o/L^2}{\rho \langle v_z \rangle^2/L} = \frac{R_o}{L} \ll 1, \tag{9.1-19}$$

indicating that pressure varies mainly in the z-direction, i.e., $\partial P/\partial r \simeq 0$ and $\partial P/\partial z = dP/dz$.

The orders of magnitude of the terms in the z-component of the equation of motion, Eq. (9.1-16), are greater than the orders of magnitude of the terms in the x-component of the equation of motion, Eq. (9.1-15), by a factor of L/R_o. Therefore, the z-component of the equation of motion is the dominant equation, expressed as

$$\underbrace{\rho \left(v_r \frac{\partial v_z}{\partial r} + v_z \frac{\partial v_z}{\partial z} \right)}_{O(\rho \langle v_z \rangle^2/L)} = -\frac{dP}{dz} + \underbrace{\frac{\mu}{r} \frac{\partial}{\partial r} \left(r \frac{\partial v_z}{\partial r} \right)}_{O(\mu \langle v_z \rangle / R_o^2)}. \tag{9.1-20}$$

If we assume that the inertial terms are negligible compared with the viscous terms, then this condition requires the following constraint:

$$O\left(\frac{\rho\langle v_z\rangle^2}{L}\right) \ll O\left(\frac{\mu\langle v_z\rangle}{R_o^2}\right) \quad \Rightarrow \quad \underbrace{\frac{\rho\langle v_z\rangle R_o}{\mu}}_{\text{Re}}\left(\frac{R_o}{L}\right) \ll 1. \qquad (9.1\text{-}21)$$

Therefore, Eq. (9.1-20) simplifies to

$$\frac{\mu}{r}\frac{\partial}{\partial r}\left(r\frac{\partial v_z}{\partial r}\right) = \frac{d\mathcal{P}}{dz}. \qquad (9.1\text{-}22)$$

In summary, Eq. (9.1-22) and $\mathcal{P} = \mathcal{P}(z)$ constitute the basic equations for the lubrication approximation. The term $d\mathcal{P}/dz$ need not be a constant and v_z may depend on z as well as on r.

Integration of Eq. (9.1-22) twice with respect to r leads to

$$v_z = \frac{1}{4\mu}\left(\frac{d\mathcal{P}}{dz}\right)r^2 + C_1(z)\ln r + C_2(z). \qquad (9.1\text{-}23)$$

Since v_z is finite at $r = 0$, then C_1 must be zero. The use of the boundary condition

$$\text{at } r = R(z) \qquad v_z = 0, \qquad (9.1\text{-}24)$$

gives the velocity distribution as

$$v_z = -\frac{R^2(z)}{4\mu}\left(\frac{d\mathcal{P}}{dz}\right)\left\{1 - \left[\frac{r}{R(z)}\right]^2\right\}. \qquad (9.1\text{-}25)$$

This is simply the equation describing the velocity distribution in a circular tube with an imposed pressure gradient, i.e., Eq. (9.1-1). In the lubrication approximation, Eq. (9.1-1) is assumed to hold locally and the term $(\mathcal{P}_o - \mathcal{P}_L)/L$ is simply replaced by $-(d\mathcal{P}/dz)$. Moreover, the radius of the tube, R, is dependent on z. One should keep in mind that the pressure gradient is an unknown quantity in Eq. (9.1-25).

The volumetric flow rate is given by

$$Q = \int_0^{2\pi}\int_0^{R(z)} v_z r\, dr\, d\theta. \qquad (9.1\text{-}26)$$

Substitution of Eq. (9.1-25) into Eq. (9.1-26) and integration give

$$Q = -\frac{\pi R^4(z)}{8\mu}\left(\frac{d\mathcal{P}}{dz}\right). \qquad (9.1\text{-}27)$$

If Eq. (9.1-2) is assumed to hold locally, then the terms $(\mathcal{P}_o - \mathcal{P}_L)/L$ and R are replaced by $-(d\mathcal{P}/dz)$ and $R(z)$, respectively, and the result is Eq. (9.1-27). In fact, Eq. (9.1-27) is considered the starting point for the lubrication approximation.

Rearrangement of Eq. (9.1-27) leads to

$$-\int_{\mathcal{P}_o}^{\mathcal{P}_L} d\mathcal{P} = \frac{8\mu Q}{\pi} \int_0^L \frac{dz}{R^4(z)}. \tag{9.1-28}$$

The variation in R with the axial distance is expressed as

$$R(z) = R_o - \epsilon z, \tag{9.1-29}$$

where

$$\epsilon = \frac{R_o - R_L}{L}. \tag{9.1-30}$$

Substitution of Eq. (9.1-29) into Eq. (9.1-28) and integration give

$$\mathcal{P}_o - \mathcal{P}_L = \frac{8\mu L Q}{3\pi} \left(\frac{R_L^{-3} - R_o^{-3}}{R_o - R_L} \right). \tag{9.1-31}$$

9.2   Flow of a Power-Law Fluid in a Tapered Tube

Consider flow of a power-law fluid in a horizontal tapered tube as shown in Figure 9.1 and determine the relationship between volumetric flow rate and pressure drop using the lubrication approximation. For a power-law fluid, Eqs. (7.4-24) and (7.4-25) give the velocity distribution and volumetric flow rate, respectively, as

$$v_z = \frac{1}{s+1} \left(\frac{\mathcal{P}_o - \mathcal{P}_L}{2mL} \right)^s R^{s+1} \left[1 - \left(\frac{r}{R} \right)^{s+1} \right], \tag{9.2-1}$$

$$Q = \frac{\pi R^{s+3}}{s+3} \left(\frac{\mathcal{P}_o - \mathcal{P}_L}{2mL} \right)^s. \tag{9.2-2}$$

If the taper angle β is small, Eq. (9.2-2) can be applied locally by replacing R, a constant, by $R(z)$ and $(\mathcal{P}_o - \mathcal{P}_L)/L$ by $-d\mathcal{P}/dz$. Thus, Eq. (9.2-2) becomes

$$Q = \frac{\pi [R(z)]^{s+3}}{s+3} \left(-\frac{1}{2m} \frac{d\mathcal{P}}{dz} \right)^s. \tag{9.2-3}$$

Substitution of Eq. (9.1-29) into Eq. (9.2-3) and rearrangement yield

$$-\int_{\mathcal{P}_o}^{\mathcal{P}_L} d\mathcal{P} = 2m \left[\frac{Q(s+3)}{\pi}\right]^{1/s} \int_0^L \frac{dz}{(R_o - \epsilon z)^{1+(3/s)}}. \tag{9.2-4}$$

Integration of Eq. (9.2-4) leads to

$$\mathcal{P}_o - \mathcal{P}_L = \frac{2msL}{3} \left[\frac{Q(s+3)}{\pi}\right]^{1/s} \left(\frac{R_L^{-3/s} - R_o^{-3/s}}{R_o - R_L}\right). \tag{9.2-5}$$

For a Newtonian fluid, i.e., $m = \mu$ and $n = s = 1$, Eq. (9.2-5) reduces to Eq. (9.1-31).

9.3   Generalized Couette Flow Between Converging Plates

Generalized Couette flow between parallel plates was covered in Section 6.1 and the volumetric flow rate is given by Eq. (6.1-8) in the form

$$Q = \frac{(\mathcal{P}_o - \mathcal{P}_L)WH^3}{12\mu L} + \frac{HWV}{2}. \tag{9.3-1}$$

Now consider the generalized Couette flow between converging planes under steady conditions as shown in Figure 9.2. When $H_o/L \ll 1$ and $(H_o - H_L)/L \ll 1$, application of the lubrication approximation assumes that Eq. (9.3-1) holds locally. Thus, replacing the terms $(\mathcal{P}_o - \mathcal{P}_L)/L$ by $-(d\mathcal{P}/dz)$ and H by $H(z)$ in Eq. (9.3-1) and rearrangement give

$$-\int_{\mathcal{P}_o}^{\mathcal{P}_L} d\mathcal{P} = \frac{12\mu}{W}\left[Q \int_0^L \frac{dz}{H^3(z)} - \frac{WV}{2}\int_0^L \frac{dz}{H^2(z)}\right]. \tag{9.3-2}$$

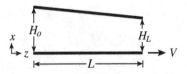

Figure 9.2   Generalized Couette flow between converging plates.

The variation in H with the axial distance is expressed as

$$H(z) = H_o - \epsilon z, \qquad (9.3\text{-}3)$$

where

$$\epsilon = \frac{H_o - H_L}{L}. \qquad (9.3\text{-}4)$$

Substitution of Eq. (9.3-3) into Eq. (9.3-2) and integration lead to

$$\mathcal{P}_o - \mathcal{P}_L = \frac{6\mu L}{H_o - H_L} \left[\frac{Q}{W} \left(\frac{1}{H_L^2} - \frac{1}{H_o^2} \right) - V \left(\frac{1}{H_L} - \frac{1}{H_o} \right) \right]. \qquad (9.3\text{-}5)$$

Chapter 10

Flow in Microchannels

The characteristic dimension of a microchannel is in the order of $1-1000\,\mu m$. Having a small volume and high surface-to-volume ratio, microchannels provide higher rates of heat and mass transfer. The study of fluid flow in microchannels is referred to as microfluidics.

10.1 Preliminaries

In gases, the mean free path, λ, is the average distance traveled by a molecule between two consecutive collisions, given by

$$\lambda = \frac{k_B T}{\sqrt{2}\pi D^2 P},$$

(10.1-1)

where k_B is the Boltzmann constant $(1.380649 \times 10^{-23}\text{ J}/\text{ K})$ and D is the molecular diameter.

The Knudsen number, Kn, is defined by

$$\text{Kn} = \frac{\lambda}{L_{\text{ch}}},$$

(10.1-2)

where L_{ch} is the characteristic length. A flow is classified into four flow regimes depending on the value of Kn:

(1) **Continuum flow** (Kn < 10^{-3}): In this case, gas–gas collisions are dominant and molecules drag one another in the direction of flow. The equation of motion with no-slip boundary conditions is used to determine the velocity field.
(2) **Slip flow** (10^{-3} < Kn < 10^{-1}): The equation of motion with slip boundary conditions is used to determine the velocity field.

(3) **Transition flow** ($10^{-1}\,\mathrm{Kn} < 10$)**:** The equation of motion is not applicable.

(4) **Free molecular flow** ($\mathrm{Kn} > 10$)**:** In this case, gas–wall collisions are dominant and molecules move independently of one another.

In the absence of temperature gradients, the slip boundary condition at a solid–fluid interface is expressed as

$$v_{\text{fluid}} - v_{\text{solid}} = -\frac{2-F}{F}\,\lambda\,\frac{\partial v}{\partial n}\bigg|_{\text{solid}}, \qquad (10.1\text{-}3)$$

where F is *Maxwell's reflection coefficient*[1] and $\partial v/\partial n$ is the velocity gradient normal to the interface.

10.2   Couette Flow of a Gas Between Parallel Plates

Consider Couette flow of an ideal gas as shown in Figure 10.1 with the following assumptions:

- steady laminar flow,
- $10^{-3} < \mathrm{Kn} < 10^{-1}$,
- negligible gravitational effects,
- constant physical properties,
- $F = 1$, i.e., the gas is fully accommodated at the wall.

Postulating $v_x = v_y = 0$ and $v_z = v_z(x)$, the z-component of the equation of motion, Eq. (F) in Table 1.5, reduces to

$$\frac{d^2 v_z}{dx^2} = 0. \qquad (10.2\text{-}1)$$

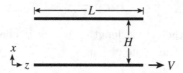

Figure 10.1   Couette flow between parallel plates.

---

[1]The value of F varies between 0 and 1. Since F is related to the tangential momentum of the incident and reflected molecules, it is also called a *tangential momentum accommodation coefficient*.

Integration of Eq. (10.2-1) twice gives

$$v_z = C_1 x + C_2. \qquad (10.2\text{-}2)$$

Using Eq. (10.1-3), the boundary conditions are expressed as

$$\text{at } x = 0 \quad v_z = V - \lambda \frac{dv_z}{dx}, \qquad (10.2\text{-}3)$$

$$\text{at } x = H \quad v_z = \lambda \frac{dv_z}{dx}. \qquad (10.2\text{-}4)$$

Therefore, the velocity distribution becomes

$$\frac{v_z}{V} = \frac{x + \lambda - H}{2\lambda - H}. \qquad (10.2\text{-}5)$$

Introduction of the dimensionless quantities

$$\xi = \frac{x}{H} \quad \text{and} \quad \text{Kn} = \frac{\lambda}{H}, \qquad (10.2\text{-}6)$$

reduces Eq. (10.2-5) to

$$\frac{v_z}{V} = \frac{\xi + \text{Kn} - 1}{2\text{Kn} - 1}. \qquad (10.2\text{-}7)$$

The volumetric flow rate is

$$Q = \int_0^W \int_0^H v_z \, dx \, dy = HW \int_0^1 v_z \, d\xi. \qquad (10.2\text{-}8)$$

Substitution of Eq. (10.2-7) into Eq. (10.2-8) and integration lead to

$$Q = \frac{HWV}{2}. \qquad (10.2\text{-}9)$$

Note that Eq. (10.2-9) is identical with Eq. (4.1-11), the volumetric flow rate between parallel plates with no-slip boundary conditions.

10.3 Flow of a Gas in a Cylindrical Microtube

Consider flow of an ideal gas in a cylindrical microtube of radius R with the following assumptions:

- steady laminar flow,
- $10^{-3} < \text{Kn} < 10^{-1}$,
- negligible gravitational effects,

- constant physical properties,
- inertial terms are negligible,
- dominant viscous force is in the r-direction,
- $F = 1$, i.e., the gas is fully accommodated at the wall.

In this case, contrary to the flow in macrochannels, the axial pressure gradient is not linear, the r-component of the velocity does not vanish, and the z-component of the velocity changes in the axial direction. The z-component of the equation of motion, Eq. (F) in Table 1.6, simplifies to

$$0 = -\frac{dP}{dz} + \frac{\mu}{r}\frac{\partial}{\partial r}\left(r\frac{\partial v_z}{\partial r}\right). \tag{10.3-1}$$

Integration of Eq. (10.3-1) with respect to r gives

$$\frac{\partial v_z}{\partial r} = \frac{1}{2\mu}\frac{dP}{dz}r + \frac{C_1}{r}. \tag{10.3-2}$$

The boundary condition at the tube center, i.e.,

$$\text{at } r = 0 \qquad \frac{\partial v_z}{\partial r} = 0, \tag{10.3-3}$$

implies that $C_1 = 0$. Integration of Eq. (10.3-2) once more results in

$$v_z = \frac{1}{4\mu}\frac{dP}{dz}r^2 + C_2. \tag{10.3-4}$$

Using Eq. (10.1-3), the boundary condition is expressed as

$$\text{at } r = R \qquad v_z = -\lambda\frac{\partial v_z}{\partial r}. \tag{10.3-5}$$

Thus, the velocity distribution is expressed in the following form:

$$v_z = -\frac{1}{4\mu}\frac{dP}{dz}R^2\left[1 - \left(\frac{r}{R}\right)^2 + \frac{2\lambda}{R}\right]. \tag{10.3-6}$$

Introduction of the dimensionless quantities

$$\xi = \frac{r}{R} \qquad \text{and} \qquad \text{Kn} = \frac{\lambda}{2R}, \tag{10.3-7}$$

reduces Eq. (10.3-6) to

$$v_z = -\frac{1}{4\mu}\frac{dP}{dz}R^2\left(1 - \xi^2 + 4\text{Kn}\right). \tag{10.3-8}$$

The equation of continuity, Eq. (B) in Table 1.1, reduces to

$$\frac{1}{r}\frac{\partial}{\partial r}(\rho v_r) + \frac{\partial}{\partial z}(\rho v_z) = 0.$$ (10.3-9)

In terms of the dimensionless distance in the r-direction, Eq. (10.3-9) becomes

$$\frac{1}{\xi}\frac{\partial}{\partial \xi}(\rho v_r) + R^2\frac{\partial}{\partial z}(\rho v_z) = 0.$$ (10.3-10)

For an ideal gas,

$$\rho = \frac{PM}{RT}.$$ (10.3-11)

Substitution of Eqs. (10.3-8) and (10.3-11) into Eq. (10.3-10) gives

$$\frac{1}{\xi}\frac{\partial}{\partial \xi}(Pv_r) - \frac{R^4}{4\mu}\frac{\partial}{\partial z}\left[P\frac{dP}{dz}(1 - \xi^2 + 4\text{Kn})\right] = 0.$$ (10.3-12)

Multiplication of Eq. (10.3-12) by $\xi\,d\xi$ and integration from $\xi = 0$ to $\xi = 1$ yield

$$\underbrace{\int_0^1 d(Pv_r)}_{0} - \frac{R^4}{4\mu}\frac{d}{dz}\left[P\frac{dP}{dz}\int_0^1 (1 - \xi^2 + 4\text{Kn})\xi\,d\xi\right] = 0.$$ (10.3-13)

Evaluation of the integral in Eq. (10.3-13) leads to

$$\frac{d}{dz}\left[(1 + 8\text{Kn})P\frac{dP}{dz}\right] = 0.$$ (10.3-14)

Therefore,

$$(1 + 8\text{Kn})P\frac{dP}{dz} = C_3.$$ (10.3-15)

The Knudsen number is

$$\text{Kn} = \frac{\lambda}{D}.$$ (10.3-16)

Substitution of Eq. (10.1-1) into Eq. (10.3-16) yields

$$\text{Kn} = \frac{\Theta}{P},$$ (10.3-17)

where

$$\Theta = \frac{k_B T}{\sqrt{2}\pi D^3}. \tag{10.3-18}$$

Substitution of Eq. (10.3-17) into Eq. (10.3-15) gives

$$\frac{1}{2}\frac{dP^2}{dz} + 8\Theta \frac{dP}{dz} = C_3. \tag{10.3-19}$$

or

$$\frac{d}{dz}\left(\frac{P^2}{2} + 8\Theta P\right) = C_3 \quad \Rightarrow \quad \frac{P^2}{2} + 8\Theta P = C_3 z + C_4. \tag{10.3-20}$$

The solution of the quadratic equation is

$$P = -8\Theta + \sqrt{64\Theta^2 + 2C_3 z + 2C_4}. \tag{10.3-21}$$

Let P_L be the value of pressure at $z = L$. Multiplication of Eq. (10.3-21) by $1/P_L$ leads to

$$\frac{P}{P_L} = -8\text{Kn}_L + \sqrt{64\text{Kn}_L^2 + C_5 z + C_6}. \tag{10.3-22}$$

where

$$\text{Kn}_L = \frac{\Theta}{P_L} \qquad C_5 = \frac{2C_3}{P_L^2} \qquad C_6 = \frac{2C_4}{P_L^2}. \tag{10.3-23}$$

Application of the boundary conditions

$$\text{at } z = 0 \qquad P = P_o, \tag{10.3-24}$$

$$\text{at } z = L \qquad P = P_L, \tag{10.3-25}$$

gives the values of C_5 and C_6 as

$$C_5 = \frac{1}{L}\left[1 - \left(\frac{P_o}{P_L}\right)^2 + 16\text{Kn}_L\left(1 - \frac{P_o}{P_L}\right)\right], \tag{10.3-26}$$

$$C_6 = \left(\frac{P_o}{P_L} + 8\text{Kn}_L\right)^2 - 64\text{Kn}_L^2. \tag{10.3-27}$$

Therefore, the pressure distribution is expressed in the following form:

$$\frac{P}{P_L} = -8\text{Kn}_L$$

$$+ \sqrt{\left(\frac{P_o}{P_L} + 8\text{Kn}_L\right)^2 + \left[1 - \left(\frac{P_o}{P_L}\right)^2 + 16\text{Kn}_L\left(1 - \frac{P_o}{P_L}\right)\right]\frac{z}{L}}. \tag{10.3-28}$$

Chapter 11

Order of Magnitude Analysis

For two- and three-dimensional flows, solution of the equations of continuity and motion becomes very complex. In these cases, the application of order of magnitude analysis provides insight into the functional form of the solution without unnecessary mathematical vigor.

11.1 Boundary Layer Over a Flat Plate

The boundary layer is the region adjacent to a solid surface that is in contact with a flowing fluid. Viscous forces cannot be neglected within the boundary layer. The effect of viscosity is only negligible in the region outside of the boundary layer. A boundary layer may be laminar or turbulent.

Consider flow of a fluid over a flat plate of length L and width W as shown in Figure 11.1. The velocity of the fluid far away from the plate, i.e., free-stream velocity, is uniform at v_∞. Velocity variation takes place within the boundary layer of thickness δ, which is dependent on the axial coordinate z.

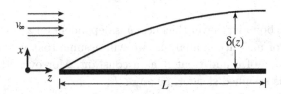

Figure 11.1 Laminar boundary layer over a flat plate.

For a laminar boundary layer, the equation of continuity, Eq. (A) in Table 1.1, and the x- and z-components of the equation of motion, Eqs. (D) and (F) in Table 1.5, are expressed as follows:

$$\frac{\partial v_x}{\partial x} + \frac{\partial v_z}{\partial z} = 0, \tag{11.1-1}$$

$$v_x \frac{\partial v_x}{\partial x} + v_z \frac{\partial v_x}{\partial z} = -\frac{1}{\rho}\frac{\partial \mathcal{P}}{\partial x} + \nu \left(\frac{\partial^2 v_x}{\partial x^2} + \frac{\partial^2 v_x}{\partial z^2} \right), \tag{11.1-2}$$

$$v_x \frac{\partial v_z}{\partial x} + v_z \frac{\partial v_z}{\partial z} = -\frac{1}{\rho}\frac{\partial \mathcal{P}}{\partial z} + \nu \left(\frac{\partial^2 v_z}{\partial x^2} + \frac{\partial^2 v_z}{\partial z^2} \right). \tag{11.1-3}$$

The modified pressure in Eqs. (11.1-2) and (11.1-3) is defined by $\mathcal{P} = P + \rho g x$. The boundary conditions are expressed as

$$\text{at } x = 0 \qquad v_x = v_z = 0, \tag{11.1-4}$$
$$\text{at } x = \delta \qquad v_z = v_\infty. \tag{11.1-5}$$

There are two characteristic lengths in this problem: the characteristic length in the x-direction is the boundary layer thickness δ over which v_z changes from 0 to v_∞ and the characteristic length in the z-direction is the length of the plate L over which changes occur in the direction of mean flow. Therefore, we can write

$$O(x) \sim \delta \qquad \text{and} \qquad O(z) \sim L. \tag{11.1-6}$$

The characteristic velocity in the z-direction is the free-stream velocity and we can propose

$$O(v_z) \sim v_\infty. \tag{11.1-7}$$

Note that the boundary layer thickness δ is dependent on z. For the purpose of the order of magnitude analysis, we will assume that we are using the maximum value of δ and treat it as a constant. Moreover, the boundary layer region is assumed slender, i.e., $\delta \ll L$.

Let U be the characteristic velocity in the x-direction. We can estimate the order of magnitude of U by using the equation of continuity,

Eq. (11.1-1). The velocity gradients in the equation of continuity have the following orders of magnitude:

$$O\left(\frac{\partial v_x}{\partial x}\right) \sim \frac{U}{\delta}, \tag{11.1-8}$$

$$O\left(\frac{\partial v_z}{\partial z}\right) \sim \frac{v_\infty}{L}. \tag{11.1-9}$$

If these two terms are retained in the equation of continuity, they must have the same order of magnitude, i.e.,

$$\frac{U}{\delta} \sim \frac{v_\infty}{L} \quad \Rightarrow \quad U \sim v_\infty \frac{\delta}{L}. \tag{11.1-10}$$

Since $\delta \ll L$, Eq. (11.1-10) implies that $U \ll v_\infty$.

In the x-component of the equation of motion, Eq. (11.1-2), the orders of magnitude of the viscous terms are

$$O\left(\frac{\partial^2 v_x}{\partial x^2}\right) \sim \frac{v_\infty}{L\delta} \quad \text{and} \quad O\left(\frac{\partial^2 v_x}{\partial z^2}\right) \sim \frac{v_\infty \delta}{L^3}, \tag{11.1-11}$$

or

$$\frac{O(\partial^2 v_x/\partial x^2)}{O(\partial^2 v_x/\partial z^2)} \sim \left(\frac{L}{\delta}\right)^2 \gg 1. \tag{11.1-12}$$

Equation (11.1-12) implies that the term $\partial^2 v_x/\partial z^2$ is much smaller than the term $\partial^2 v_x/\partial x^2$ and can be neglected.

Similarly, in the z-component of the equation of motion, Eq. (11.1-3), the order of magnitude analysis leads to

$$O\left(\frac{\partial^2 v_z}{\partial x^2}\right) \sim \frac{v_\infty}{\delta^2} \quad \text{and} \quad O\left(\frac{\partial^2 v_z}{\partial z^2}\right) \sim \frac{v_\infty}{L^2}, \tag{11.1-13}$$

or

$$\frac{\partial^2 v_z/\partial x^2}{\partial^2 v_z/\partial z^2} \sim \left(\frac{L}{\delta}\right)^2 \gg 1. \tag{11.1-14}$$

Therefore, $\partial^2 v_z/\partial z^2$ is much smaller than the term $\partial^2 v_z/\partial x^2$ and can also be neglected.

With these simplifications, Eqs. (11.1-2) and (11.1-3) reduce to

$$\underbrace{v_x \frac{\partial v_x}{\partial x} + v_z \frac{\partial v_x}{\partial z}}_{O(v_\infty^2 \delta/L^2)} = -\frac{1}{\rho}\frac{\partial P}{\partial x} + \underbrace{\nu \frac{\partial^2 v_x}{\partial x^2}}_{O(\nu v_\infty/L\delta)}, \tag{11.1-15}$$

$$v_x \frac{\partial v_z}{\partial x} + v_z \frac{\partial v_z}{\partial z} = -\frac{1}{\rho} \frac{\partial \mathcal{P}}{\partial z} + \nu \frac{\partial^2 v_z}{\partial x^2} . \tag{11.1-16}$$
$$\underbrace{\hphantom{v_x \frac{\partial v_z}{\partial x} + v_z \frac{\partial v_z}{\partial z}}}_{O(v_\infty^2/L)} \qquad\qquad \underbrace{\hphantom{\nu \frac{\partial^2 v_z}{\partial x^2}}}_{O(\nu v_\infty/\delta^2)}$$

The term ν in Eqs. (11.1-15) and (11.1-16) is the kinematic viscosity, defined by μ/ρ. Equations (11.1-15) and (11.1-16) consist of inertial, pressure, and viscous terms and they must be of comparable magnitude. Consequently, either the orders of magnitude of the pressure and inertial terms or the pressure and viscous terms must be equal to each other. The order of magnitude of the pressure term in Eq. (11.1-15) is

$$O\left(\frac{1}{\rho}\frac{\partial \mathcal{P}}{\partial x}\right) \sim \frac{v_\infty^2 \delta}{L^2} \qquad \text{or} \qquad O\left(\frac{1}{\rho}\frac{\partial \mathcal{P}}{\partial x}\right) \sim \frac{\nu v_\infty}{L\delta}. \tag{11.1-17}$$

On the other hand, the order of magnitude of the pressure term in Eq. (11.1-16) is

$$O\left(\frac{1}{\rho}\frac{\partial \mathcal{P}}{\partial z}\right) \sim \frac{v_\infty^2}{L} \qquad \text{or} \qquad O\left(\frac{1}{\rho}\frac{\partial \mathcal{P}}{\partial z}\right) \sim \frac{\nu v_\infty}{\delta^2}. \tag{11.1-18}$$

Therefore,

$$\frac{\partial \mathcal{P}/\partial x}{\partial \mathcal{P}/\partial z} \sim \frac{\rho v_\infty^2 \delta/L^2}{\rho v_\infty^2/L} = \frac{\delta}{L} \ll 1, \tag{11.1-19}$$

indicating that pressure varies mainly in the z-direction, i.e., $\partial \mathcal{P}/\partial x \simeq 0$ and $\partial \mathcal{P}/\partial z = d\mathcal{P}/dz$. Since $\mathcal{P} = \mathcal{P}_\infty$ at $x = \delta$, then

$$\frac{\partial \mathcal{P}}{\partial z} = \frac{d\mathcal{P}}{dz} = \frac{d\mathcal{P}_\infty}{dz}. \tag{11.1-20}$$

The orders of magnitude of the terms in the z-component of the equation of motion, Eq. (11.1-16), are greater than the orders of magnitude of the terms in the x-component of the equation of motion, Eq. (11.1-15), by a factor of L/δ. Therefore, the z-component of the equation of motion is the dominant equation within the boundary layer, expressed as

$$v_x \frac{\partial v_z}{\partial x} + v_z \frac{\partial v_z}{\partial z} = -\frac{1}{\rho} \frac{\partial \mathcal{P}_\infty}{\partial z} + \nu \frac{\partial^2 v_z}{\partial x^2} . \tag{11.1-21}$$
$$\underbrace{\hphantom{v_x \frac{\partial v_z}{\partial x} + v_z \frac{\partial v_z}{\partial z}}}_{O(v_\infty^2/L)} \qquad\qquad \underbrace{\hphantom{\nu \frac{\partial^2 v_z}{\partial x^2}}}_{O(\nu v_\infty/\delta^2)}$$

Since the inertial and viscous terms have comparable magnitudes, then

$$O\left(\frac{v_\infty^2}{L}\right) = O\left(\frac{\nu v_\infty}{\delta^2}\right) \qquad \Rightarrow \qquad O(\delta) \sim \frac{L\nu}{v_\infty}. \tag{11.1-22}$$

If the Reynolds number based on the length of the plate is defined by

$$\text{Re}_L = \frac{L v_\infty}{\nu},$$ (11.1-23)

then Eq. (11.1-22) takes the form of

$$O\left(\frac{\delta}{L}\right) \sim \text{Re}_L^{-1/2}.$$ (11.1-24)

The analysis presented above is valid when $\text{Re}_L < 500,000$, i.e., laminar boundary layer.

11.1.1 *Calculation of drag force*

The drag force exerted by the fluid on the plate surface is given by Eq. (4.1-15), in which $\boldsymbol{\lambda} = \mathbf{e}_z$ and $\mathbf{n}$, the unit normal vector directed from the fluid to the plate surface, is $-\mathbf{e}_x$. Since $\boldsymbol{\lambda}$ is orthogonal to $\mathbf{n}$, form drag is zero and Eq. (4.1-15) simplifies to

$$F_D = -\int_0^L \int_0^W \tau_{xz}\big|_{x=0} \, dy \, dz = \int_0^L \int_0^W \mu \frac{\partial v_z}{\partial x}\bigg|_{x=0} \, dy \, dz.$$ (11.1-25)

The order of magnitude of the shear stress on the plate surface is

$$O\left(\mu \frac{\partial v_z}{\partial x}\bigg|_{x=0}\right) \sim \frac{\mu v_\infty}{\delta}.$$ (11.1-26)

Multiplication of the numerator and denominator of Eq. (11.1-26) by L yields

$$O\left(\mu \frac{\partial v_z}{\partial x}\bigg|_{x=0}\right) \sim \frac{\mu v_\infty}{L} \frac{L}{\delta}.$$ (11.1-27)

The use of Eq. (11.1-24) in Eq. (11.1-27) gives

$$O\left(\mu \frac{\partial v_z}{\partial x}\bigg|_{x=0}\right) \sim \frac{\mu v_\infty}{L} \text{Re}_L^{1/2}.$$ (11.1-28)

From the definition of the Reynolds number, given by Eq. (11.1-23),

$$L = \frac{\mu \text{Re}_L}{\rho v_\infty}.$$ (11.1-29)

Substitution of Eq. (11.1-29) into Eq. (11.1-28) leads to

$$O\left(\mu \frac{\partial v_z}{\partial x}\bigg|_{x=0}\right) \sim \rho v_\infty^2 \text{Re}_L^{-1/2}.$$ (11.1-30)

Therefore, the drag force has the following order of magnitude:

$$O(F_D) \sim A_{\text{ch}} \rho v_\infty^2 \text{Re}_L^{-1/2},$$ (11.1-31)

where the characteristic area, A_{ch}, is the area of the plate, i.e., WL.

11.1.2 *Friction factor*

The friction factor is defined by Eq. (5.1-24), i.e.,

$$F_D = A_{ch} K_{ch} f. \tag{11.1-32}$$

The order of magnitude of the characteristic kinetic energy is

$$O(K_{ch}) \sim \rho v_\infty^2. \tag{11.1-33}$$

Substitution of Eqs. (11.1-31) and (11.1-33) into Eq. (11.1-32) gives

$$O(f) \sim \text{Re}_L^{-1/2}. \tag{11.1-34}$$

11.2 Flow in a Circular Tube With a Permeable Wall

Consider steady flow of an incompressible liquid in a cylindrical tube of radius R and length L ($R \ll L$). The tube has a porous wall and the velocity of the liquid normal to the tube wall is given as $v_r|_{r=R} = U$. Let us assume that the leakage rate is small so that the average axial velocity, $\langle v_z \rangle$, does not change much.

Postulating $v_r = v_r(r, z)$, $v_\theta = 0$, $v_z = v_z(r, z)$, and $P = P(r, z)$, the equation of continuity, Eq. (B) in Table 1.1, and the r- and z-components of the equation of motion, Eqs. (D) and (F) in Table 1.6, take the form

$$\frac{1}{r}\frac{\partial}{\partial r}(rv_r) + \frac{\partial v_z}{\partial z} = 0, \tag{11.2-1}$$

$$\rho\left(v_r\frac{\partial v_r}{\partial r} + v_z\frac{\partial v_r}{\partial z}\right) = -\frac{\partial P}{\partial r} + \mu\left\{\frac{\partial}{\partial r}\left[\frac{1}{r}\frac{\partial}{\partial r}(rv_r)\right] + \frac{\partial^2 v_r}{\partial z^2}\right\}, \tag{11.2-2}$$

$$\rho\left(v_r\frac{\partial v_z}{\partial r} + v_z\frac{\partial v_z}{\partial z}\right) = -\frac{\partial P}{\partial z} + \mu\left[\frac{1}{r}\frac{\partial}{\partial r}\left(r\frac{\partial v_z}{\partial r}\right) + \frac{\partial^2 v_z}{\partial z^2}\right]. \tag{11.2-3}$$

Let the characteristic velocities in the r- and z-directions be U and $\langle v_z \rangle$, respectively. Thus, the velocity gradients in the equation of continuity have the following orders of magnitude:

$$O\left[\frac{1}{r}\frac{\partial}{\partial r}(rv_r)\right] \sim \frac{U}{R}, \tag{11.2-4}$$

$$O\left(\frac{\partial v_z}{\partial z}\right) \sim \frac{\langle v_z \rangle}{L}. \tag{11.2-5}$$

If these two terms are retained in the equation of continuity, they must have the same order of magnitude, i.e.,

$$\frac{U}{R} \sim \frac{\langle v_z \rangle}{L} \qquad \Rightarrow \qquad O(U) \sim \langle v_z \rangle \frac{R}{L}. \tag{11.2-6}$$

Since $R/L \ll 1$, Eq. (11.2-6) implies that $U \ll \langle v_z \rangle$.

In the r-component of the equation of motion, Eq. (11.2-2), the orders of magnitude of the viscous terms are

$$O\left\{ \mu \frac{\partial}{\partial r} \left[\frac{1}{r} \frac{\partial}{\partial r} (r v_r) \right] \right\} \sim \frac{\mu \langle v_z \rangle}{RL} \quad \text{and} \quad O\left(\mu \frac{\partial^2 v_r}{\partial z^2} \right) \sim \frac{\mu \langle v_z \rangle R}{L^3}. \tag{11.2-7}$$

or

$$\frac{O\left\{ \mu \dfrac{\partial}{\partial r} \left[\dfrac{1}{r} \dfrac{\partial}{\partial r} (r v_r) \right] \right\}}{O\left(\mu \dfrac{\partial^2 v_r}{\partial z^2} \right)} \sim \left(\frac{L}{R} \right)^2 \gg 1. \tag{11.2-8}$$

Equation (11.2-8) implies that the term $\mu(\partial^2 v_r / \partial z^2)$ in Eq. (11.2-2) is very small and can be neglected.

Similarly, in the z-component of the equation of motion, Eq. (11.2-3), the order of magnitude analysis leads to

$$O\left[\frac{\mu}{r} \frac{\partial}{\partial r} \left(r \frac{\partial v_z}{\partial r} \right) \right] \sim \frac{\mu \langle v_z \rangle}{R^2} \quad \text{and} \quad O\left(\mu \frac{\partial^2 v_z}{\partial z^2} \right) \sim \frac{\mu \langle v_z \rangle}{L^2}, \tag{11.2-9}$$

or

$$\frac{\dfrac{\mu}{r} \dfrac{\partial}{\partial r} \left(r \dfrac{\partial v_z}{\partial r} \right)}{\mu \dfrac{\partial^2 v_z}{\partial z^2}} \sim \left(\frac{L}{R} \right)^2 \gg 1. \tag{11.2-10}$$

Therefore, the term $\mu(\partial^2 v_z / \partial z^2)$ in Eq. (11.2-3) is very small and can be neglected.

With these simplifications, Eqs. (11.2-2) and (11.2-3) reduce to

$$\underbrace{v_r \frac{\partial v_r}{\partial r} + v_z \frac{\partial v_r}{\partial z}}_{O(\langle v_z \rangle^2 R / L^2)} = -\frac{1}{\rho} \frac{\partial \mathcal{P}}{\partial r} + \underbrace{\nu \frac{\partial}{\partial r} \left[\frac{1}{r} \frac{\partial}{\partial r} (r v_r) \right]}_{O(\nu \langle v_z \rangle / RL)}, \tag{11.2-11}$$

$$\underbrace{v_r \frac{\partial v_z}{\partial r} + v_z \frac{\partial v_z}{\partial z}}_{O(\langle v_z \rangle^2 / L)} = -\frac{1}{\rho} \frac{\partial \mathcal{P}}{\partial z} + \underbrace{\frac{\nu}{r} \frac{\partial}{\partial r} \left(r \frac{\partial v_z}{\partial r} \right)}_{O(\nu \langle v_z \rangle / R^2)}. \tag{11.2-12}$$

The term ν in Eqs. (11.2-11) and (11.2-12) is the kinematic viscosity, defined by μ/ρ. Equations (11.2-11) and (11.2-12) consist of inertial, pressure, and viscous terms and they must be of comparable magnitude. Consequently, either the orders of magnitude of the pressure and inertial terms or the pressure and viscous terms must be equal to each other. The order of magnitude of the pressure term in Eq. (11.2-11) is

$$O\left(\frac{1}{\rho}\frac{\partial \mathcal{P}}{\partial r}\right) \sim \frac{\langle v_z\rangle^2 R}{L^2} \quad \text{or} \quad O\left(\frac{1}{\rho}\frac{\partial \mathcal{P}}{\partial r}\right) \sim \frac{\nu\langle v_z\rangle}{RL}. \tag{11.2-13}$$

On the other hand, the order of magnitude of the pressure term in Eq. (11.2-12) is

$$O\left(\frac{1}{\rho}\frac{\partial \mathcal{P}}{\partial z}\right) \sim \frac{\langle v_z\rangle^2}{L} \quad \text{or} \quad O\left(\frac{1}{\rho}\frac{\partial \mathcal{P}}{\partial z}\right) \sim \frac{\nu\langle v_z\rangle}{R_o^2}. \tag{11.2-14}$$

Therefore,

$$\frac{\partial \mathcal{P}/\partial r}{\partial \mathcal{P}/\partial z} \sim \frac{\rho\langle v_z\rangle^2 R/L^2}{\rho\langle v_z\rangle^2/L} = \frac{R}{L} \ll 1, \tag{11.2-15}$$

indicating that pressure varies mainly in the z-direction, i.e., $\partial \mathcal{P}/\partial r \simeq 0$ and $\partial \mathcal{P}/\partial z = d\mathcal{P}/dz$.

The orders of magnitude of the terms in the z-component of the equation of motion, Eq. (11.2-12), are greater than the orders of magnitude of the terms in the r-component of the equation of motion, Eq. (11.2-11), by a factor of L/R. Therefore, the z-component of the equation of motion is the dominant equation, expressed as

$$\rho\underbrace{\left(v_r\frac{\partial v_z}{\partial r} + v_z\frac{\partial v_z}{\partial z}\right)}_{O(\rho\langle v_z\rangle^2/L)} = -\frac{d\mathcal{P}}{dz} + \underbrace{\frac{\mu}{r}\frac{\partial}{\partial r}\left(r\frac{\partial v_z}{\partial r}\right)}_{O(\mu\langle v_z\rangle/R^2)}. \tag{11.2-16}$$

When the inertial terms are negligible compared with the viscous terms, then this condition requires the following constraint:

$$O\left(\frac{\rho\langle v_z\rangle^2}{L}\right) \ll O\left(\frac{\mu\langle v_z\rangle}{R^2}\right) \quad \Rightarrow \quad \frac{\rho\langle v_z\rangle R}{\mu}\left(\frac{R}{L}\right) \ll 1. \tag{11.2-17}$$

If the constraint given by Eq. (11.2-17) is satisfied, then Eq. (11.2-12) simplifies to

$$\frac{\mu}{r}\frac{\partial}{\partial r}\left(r\frac{\partial v_z}{\partial r}\right) = \frac{d\mathcal{P}}{dz}. \tag{11.2-18}$$

Integration of Eq. (11.2-18) twice with respect to r leads to

$$v_z = \frac{1}{4\mu} \left(\frac{dP}{dz} \right) r^2 + C_1 \ln r + C_2. \tag{11.2-19}$$

Since v_z is finite at $r = 0$, then C_1 must be zero. The use of the boundary condition

$$\text{at } r = R \qquad v_z = 0 \tag{11.2-20}$$

gives the velocity distribution as

$$v_z = -\frac{R^2}{4\mu} \left(\frac{dP}{dz} \right) \left[1 - \left(\frac{r}{R} \right)^2 \right]. \tag{11.2-21}$$

To complete the analysis, it is necessary to determine the pressure gradient. For this purpose, we will consider the following two cases.

Case (i) **U is a known constant and the average axial velocity at the tube inlet, $\langle v_z \rangle_o$, is given**

The average velocity is

$$\langle v_z \rangle = \frac{1}{\pi R^2} \int_0^{2\pi} \int_0^R v_z r \, dr \, d\theta = 2 \int_0^1 v_z \xi \, d\xi, \tag{11.2-22}$$

where the dimensionless distance ξ is defined by

$$\xi = \frac{r}{R}. \tag{11.2-23}$$

Substitution of Eq. (11.2-21) into Eq. (11.2-22) and integration give

$$\langle v_z \rangle = -\frac{1}{8\mu} \left(\frac{dP}{dz} \right) R^2. \tag{11.2-24}$$

The use of Eq. (11.2-24) in Eq. (11.2-21) yields

$$v_z = 2 \langle v_z \rangle \left[1 - \left(\frac{r}{R} \right)^2 \right]. \tag{11.2-25}$$

Substitution of Eq. (11.2-25) into Eq. (11.2-1) results in

$$\frac{\partial}{\partial r} (r v_r) = -2 \frac{d \langle v_z \rangle}{dz} \left(r - \frac{r^3}{R^2} \right). \tag{11.2-26}$$

Integration of Eq. (11.2-26) with respect to r gives

$$v_r = -2 \frac{d \langle v_z \rangle}{dz} \left(\frac{r}{2} - \frac{r^3}{4R^2} \right) + \frac{C_3}{r}. \tag{11.2-27}$$

Since v_r is finite at $r = 0$, then C_3 must be zero and Eq. (11.2-27) reduces to

$$v_r = -\frac{d\langle v_z \rangle}{dz}\left(r - \frac{r^3}{2R^2}\right). \tag{11.2-28}$$

The use of the boundary condition

$$\text{at } r = R \qquad v_r = U, \tag{11.2-29}$$

simplifies Eq. (11.2-28) to

$$U = -\frac{d\langle v_z \rangle}{dz}\frac{R}{2} \quad \Rightarrow \quad \frac{d\langle v_z \rangle}{dz} = -\frac{2U}{R}. \tag{11.2-30}$$

The use of Eq. (11.2-30) in Eq. (11.2-28) leads to

$$v_r = 2U\left[\frac{r}{R} - \frac{1}{2}\left(\frac{r}{R}\right)^3\right]. \tag{11.2-31}$$

Integration of Eq. (11.2-30) with respect to z gives

$$\langle v_z \rangle = -\frac{2U}{R}z + C_4. \tag{11.2-32}$$

The boundary condition at the tube inlet is

$$\text{at } z = 0 \qquad \langle v_z \rangle = \langle v_z \rangle_o. \tag{11.2-33}$$

Therefore, Eq. (11.2-32) becomes

$$\langle v_z \rangle = \langle v_z \rangle_o - \frac{2U}{R}z. \tag{11.2-34}$$

Substitution of Eq. (11.2-34) into Eq. (11.2-25) gives the z-component of velocity as

$$v_z = 2\left(\langle v_z \rangle_o - \frac{2U}{R}z\right)\left[1 - \left(\frac{r}{R}\right)^2\right]. \tag{11.2-35}$$

To obtain the pressure distribution, let us substitute Eq. (11.2-34) into Eq. (11.2-24). The result is

$$-\int_{\mathcal{P}_o}^{\mathcal{P}_L} d\mathcal{P} = \frac{8\mu}{R^2}\int_0^L\left(\langle v_z \rangle_o - \frac{2U}{R}z\right)dz. \tag{11.2-36}$$

Integration of Eq. (11.2-36) leads to

$$\mathcal{P}_o - \mathcal{P}_L = \frac{8\mu\langle v_z \rangle_o L}{R^2}\left(1 - \frac{UL}{R\langle v_z \rangle_o}\right). \tag{11.2-37}$$

Equation (11.2-35) indicates that $v_z = 0$ when

$$\langle v_z \rangle_o = \frac{2U}{R} z \qquad \Rightarrow \qquad z = \frac{\langle v_z \rangle_o R}{2U}. \qquad (11.2\text{-}38)$$

However, the analysis presented above is valid as long as the order of magnitude $\langle v_z \rangle$ does not differ much from $\langle v_z \rangle_o$, i.e., U remains constant. From Eq. (11.2-38), the criterion

$$L \ll \frac{\langle v_z \rangle_o R}{2U}, \qquad (11.2\text{-}39)$$

must be satisfied in order to keep U constant.

Case (ii) U is determined from Darcy's law

Flow through porous media is governed by Darcy's law. Thus, the radial velocity at the tube surface is given by

$$v_r|_{r=R} = U = \frac{K}{\mu} \frac{(\mathcal{P} - \mathcal{P}_L)}{\ell}, \qquad (11.2\text{-}40)$$

where K and ℓ are the permeability and thickness of the porous wall, respectively. In writing Eq. (11.2-40), we are implicitly assuming that the pressure in the space outside the tube is constant at $\mathcal{P}_L$.

Multiplication of Eq. (11.2-1) by $r\,dr$ and integration from $r = 0$ to $r = R$ give

$$\int_0^R \frac{\partial}{\partial r}(rv_r)\,dr + \int_0^R \frac{\partial v_z}{\partial z} r\,dr = 0, \qquad (11.2\text{-}41)$$

or

$$(rv_r)|_{r=R} + \int_0^R \frac{\partial v_z}{\partial z} r\,dr = 0. \qquad (11.2\text{-}42)$$

From Eq. (11.2-21),

$$\frac{\partial v_z}{\partial z} = -\frac{R^2}{4\mu}\left(\frac{d^2\mathcal{P}}{dz^2}\right)\left[1 - \left(\frac{r}{R}\right)^2\right]. \qquad (11.2\text{-}43)$$

Substitution of Eqs. (11.2-40) and (11.2-43) into Eq. (11.2-42) yields

$$\frac{K}{\mu}\frac{(\mathcal{P} - \mathcal{P}_L)R}{\ell} - \frac{R^2}{4\mu}\left(\frac{d^2\mathcal{P}}{dz^2}\right)\int_0^R \left[1 - \left(\frac{r}{R}\right)^2\right] r\,dr = 0. \qquad (11.2\text{-}44)$$

Carrying out the integration simplifies Eq. (11.2-44) to

$$\frac{K}{\ell}(\mathcal{P} - \mathcal{P}_L) - \frac{R^3}{16}\frac{d^2\mathcal{P}}{dz^2} = 0. \qquad (11.2\text{-}45)$$

Introduction of the dimensionless quantities

$$\Theta = \frac{\mathcal{P} - \mathcal{P}_L}{\mathcal{P}_o - \mathcal{P}_L} \qquad \eta = \frac{z}{L} \qquad \beta = \sqrt{\frac{KL^2}{\ell R^3}}, \tag{11.2-46}$$

transforms Eq. (11.2-46) into

$$\frac{d^2\Theta}{d\eta^2} - 16\beta^2\Theta = 0, \tag{11.2-47}$$

which is subject to the following boundary conditions:

$$\text{at } \eta = 0 \quad \Theta = 1, \tag{11.2-48}$$
$$\text{at } \eta = 1 \quad \Theta = 0. \tag{11.2-49}$$

The solution of Eq. (11.2-47) is

$$\Theta = \frac{\sinh\left[4\beta(1 - \eta)\right]}{\sinh(4\beta)}. \tag{11.2-50}$$

Therefore, the pressure gradient is

$$\frac{d\mathcal{P}}{dz} = \left(\frac{\mathcal{P}_o - \mathcal{P}_L}{L}\right)\frac{d\Theta}{d\eta} = -4\beta\left(\frac{\mathcal{P}_o - \mathcal{P}_L}{L}\right)\frac{\cosh\left[4\beta(1 - \eta)\right]}{\sinh(4\beta)}. \tag{11.2-51}$$

Substitution of Eq. (11.2-51) into Eq. (11.2-21) gives the axial velocity distribution.

Chapter 12

Two-Dimensional Flow

12.1 Stream Function

As stated in Section 1.1.2, a streamline is a line in the flow field that is everywhere tangent to the velocity. Consequently, there can be no flow across a streamline. In other words, every streamline could be replaced by a solid boundary.

Let us consider two-dimensional flow of an incompressible fluid in the cylindrical coordinate system with the following velocity components:

$$v_r = v_r(r, \theta) \qquad v_\theta = v_\theta(r, \theta) \qquad v_z = 0. \tag{12.1-1}$$

Under these conditions, the equation of continuity, Eq. (B) in Table 1.1, simplifies to

$$\frac{\partial}{\partial r}(r v_r) + \frac{\partial v_\theta}{\partial \theta} = 0. \tag{12.1-2}$$

Partial derivatives are independent of the order of differentiation. For example, for a function ψ,

$$\frac{\partial^2 \psi}{\partial r \partial \theta} = \frac{\partial^2 \psi}{\partial \theta \partial r}. \tag{12.1-3}$$

Comparison of Eqs. (12.1-2) and (12.1-3) indicates that when the velocity components are defined in terms of a function ψ, called a *stream function*, as

$$v_r = -\frac{1}{r}\frac{\partial \psi}{\partial \theta} \qquad \text{and} \qquad v_\theta = \frac{\partial \psi}{\partial r}, \tag{12.1-4}$$

the equation of continuity, Eq. (12.1-2), is satisfied identically. Therefore, given a stream function, $\psi = \psi(r, \theta)$, the velocity at every point in the flow field can be calculated using Eq. (12.1-4).

Note that reversing the signs in Eq. (12.1-4), i.e.,

$$v_r = \frac{1}{r}\frac{\partial \psi}{\partial \theta} \quad \text{and} \quad v_\theta = -\frac{\partial \psi}{\partial r}, \tag{12.1-5}$$

also satisfies the equation of continuity. In the literature, it is possible to come across either Eq. (12.1-4) or (12.1-5) depending on the sign convention used by the authors.

In any two-dimensional flow, the equation of motion has two components. The pressure term can be eliminated between these two equations using the fact that

$$\frac{\partial^2 P}{\partial x_1 \partial x_2} = \frac{\partial^2 P}{\partial x_2 \partial x_1}. \tag{12.1-6}$$

Expressing velocity components in terms of a stream function leads to partial differential equations for ψ that are equivalent to the Navier–Stokes equations. These equations are given in Table 12.1.

12.2 Flow Between Parallel Plates

Flow of a Newtonian fluid between parallel plates was covered in Section 5.1. For the flow geometry shown in Figure 12.1, let us rework the same problem using a stream function.

Following the procedure given in Section 5.1, the velocity components are postulated as

$$v_y = v_z = 0 \quad \text{and} \quad v_x = v_x(y). \tag{12.2-1}$$

Expressing v_y in terms of ψ gives

$$v_y = \frac{\partial \psi}{\partial x} = 0 \quad \Longrightarrow \quad \psi \neq \psi(x). \tag{12.2-2}$$

Figure 12.1   Flow between parallel plates.

Table 12.1 Stream function equations.

Functional form of velocities	Velocity components	Governing equation	Differential operators
$v_x = v_x(x,y)$ $v_y = v_y(x,y)$ $v_z = 0$	$v_x = -\dfrac{\partial\psi}{\partial y}$ $v_y = \dfrac{\partial\psi}{\partial x}$	$\dfrac{\partial}{\partial t}(\nabla^2\psi) + \dfrac{\partial(\psi,\nabla^2\psi)}{\partial(x,y)} = \nu\nabla^4\psi$	$\nabla^2 = \dfrac{\partial^2}{\partial x^2} + \dfrac{\partial^2}{\partial y^2}$ $\nabla^4\psi = \nabla^2(\nabla^2\psi)$ (A)
$v_r = v_r(r,\theta)$ $v_\theta = v_\theta(r,\theta)$ $v_z = 0$	$v_r = -\dfrac{1}{r}\dfrac{\partial\psi}{\partial\theta}$ $v_\theta = \dfrac{\partial\psi}{\partial r}$	$\dfrac{\partial}{\partial t}(\nabla^2\psi) + \dfrac{1}{r}\dfrac{\partial(\psi,\nabla^2\psi)}{\partial(r,\theta)} = \nu\nabla^4\psi$	$\nabla^2 = \dfrac{\partial^2}{\partial r^2} + \dfrac{1}{r}\dfrac{\partial}{\partial r} + \dfrac{1}{r^2}\dfrac{\partial^2}{\partial\theta^2}$ (B)
$v_r = v_r(r,z)$ $v_z = v_z(r,z)$ $v_\theta = 0$	$v_r = \dfrac{1}{r}\dfrac{\partial\psi}{\partial z}$ $v_z = -\dfrac{1}{r}\dfrac{\partial\psi}{\partial r}$	$\dfrac{\partial}{\partial t}(E^2\psi) - \dfrac{1}{r}\dfrac{\partial(\psi,E^2\psi)}{\partial(r,z)} - \dfrac{2}{r^2}\dfrac{\partial\psi}{\partial z}E^2\psi$ $= \nu E^4\psi$	$E^2 = \dfrac{\partial^2}{\partial r^2} - \dfrac{1}{r}\dfrac{\partial}{\partial r} + \dfrac{\partial^2}{\partial z^2}$ $E^4\psi = E^2(E^2\psi)$ (C)
$v_r = v_r(r,\theta)$ $v_\theta = v_z(r,\theta)$ $v_\phi = 0$	$v_r = -\dfrac{1}{r^2\sin\theta}\dfrac{\partial\psi}{\partial\theta}$ $v_\theta = \dfrac{1}{r\sin\theta}\dfrac{\partial\psi}{\partial r}$	$\dfrac{\partial}{\partial t}(E^2\psi) + \dfrac{1}{r^2\sin\theta}\dfrac{\partial(\psi,E^2\psi)}{\partial(r,\theta)} - \dfrac{2E^2\psi}{r^2\sin^2\theta}$ $\times\left(\dfrac{\partial\psi}{\partial r}\cos\theta - \dfrac{1}{r}\dfrac{\partial\psi}{\partial\theta}\sin\theta\right) = \nu E^4\psi$	$E^2 = \dfrac{\partial^2}{\partial r^2} + \dfrac{\sin\theta}{r^2}\dfrac{\partial}{\partial\theta}\left(\dfrac{1}{\sin\theta}\dfrac{\partial}{\partial\theta}\right)$ (D)

Since the stream function depends only on the y-coordinate, the biharmonic equation, Eq. (A) in Table 12.1, reduces to

$$\nabla^4 \psi = \nabla^2(\nabla^2 \psi) = \frac{d^4 \psi}{dy^4} = 0. \tag{12.2-3}$$

The solution of Eq. (12.2-3) is given by

$$\psi = C_1 + C_2\, y + C_3\, y^2 + C_4\, y^3. \tag{12.2-4}$$

Four boundary conditions are needed to evaluate the integration constants.

Any solid wall is a streamline and therefore has a constant value of ψ. Since ψ is arbitrary to a constant, we can fix the value on one wall to any value we choose. Thus, the first boundary condition can be specified as

$$\text{at } y = 0 \qquad \psi = 0. \tag{12.2-5}$$

The no-slip boundary conditions on the wall surfaces are expressed as

$$\text{at } y = 0 \quad v_x = 0 \qquad \Rightarrow \qquad \frac{d\psi}{dy} = 0, \tag{12.2-6}$$

$$\text{at } y = H \quad v_x = 0 \qquad \Rightarrow \qquad \frac{d\psi}{dy} = 0. \tag{12.2-7}$$

The volumetric flow rate is given by

$$\mathcal{Q} = \int_0^W \int_0^H v_x\, dy\, dz = W \int_0^H v_x\, dy = -W \int_0^H \frac{d\psi}{dy}\, dy$$
$$= W \left(\psi|_{y=0} - \psi|_{y=H} \right) = -W\, \psi|_{y=H}. \tag{12.2-8}$$

Therefore, the last boundary condition is

$$\text{at } y = H \qquad \psi = -\frac{\mathcal{Q}}{W}. \tag{12.2-9}$$

Application of Eqs. (12.2-5) and (12.2-6) gives $C_1 = 0$ and $C_2 = 0$, respectively. Application of Eqs. (12.2-7) and (12.2-9) gives the stream function as

$$\psi = \frac{\mathcal{Q}}{W} \left[2 \left(\frac{y}{H} \right)^3 - 3 \left(\frac{y}{H} \right)^2 \right]. \tag{12.2-10}$$

Therefore, the component of velocity in the x-direction is

$$v_x = -\frac{d\psi}{dy} = \frac{6\mathcal{Q}}{WH} \left[\frac{y}{H} - \left(\frac{y}{H} \right)^2 \right]. \tag{12.2-11}$$

When the stream function is used in the problem solution, it naturally brings in the flow rate $\mathcal{Q}$ as a boundary condition and information about pressure gradient is not needed.

For flow between parallel plates, the equation of motion was given by Eq. (5.1-9). For the coordinate system shown in Figure 12.1, it is expressed as

$$\mu \frac{d^2 v_x}{dy^2} = \frac{d\mathcal{P}}{dx} = -\frac{\mathcal{P}_o - \mathcal{P}_L}{L}. \tag{12.2-12}$$

The use of Eq. (12.2-11) in Eq. (12.2-12) results in

$$-\frac{12\mu\mathcal{Q}}{WH^3} = -\frac{\mathcal{P}_o - \mathcal{P}_L}{L} \quad \Rightarrow \quad \mathcal{Q} = \frac{(\mathcal{P}_o - \mathcal{P}_L)WH^3}{12\mu L}, \tag{12.2-13}$$

which is identical with Eq. (5.1-18).

12.3 Creeping Flow Around a Sphere

A Newtonian fluid is flowing toward a stationary sphere of radius R with velocity v_∞, where v_∞ is small enough for the Reynolds number to be much less than one, i.e., $\mathrm{Re} = 2Rv_\infty\rho/\mu \ll 1$. Therefore, the left-hand side of Eq. (D) in Table 12.1 is zero and the governing equation for ψ reduces to

$$E^4\psi = 0, \tag{12.3-1}$$

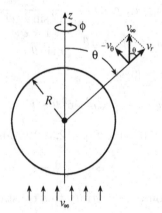

Figure 12.2   Creeping flow around a sphere.

or

$$\left[\frac{\partial^2}{\partial r^2} + \frac{\sin\theta}{r^2}\frac{\partial}{\partial\theta}\left(\frac{1}{\sin\theta}\frac{\partial}{\partial\theta}\right)\right]^2 \psi = 0. \tag{12.3-2}$$

The solution of Eq. (12.3-2) requires four boundary conditions.

The velocity components in terms of the stream function are expressed as

$$v_r = -\frac{1}{r^2\sin\theta}\frac{\partial\psi}{\partial\theta}, \tag{12.3-3}$$

$$v_\theta = \frac{1}{r\sin\theta}\frac{\partial\psi}{\partial r}. \tag{12.3-4}$$

Since $v_r = v_\theta = 0$ on the surface of the sphere, then it follows from Eqs. (12.3-3) and (12.3-4) that

$$\text{at } r = R \qquad \frac{\partial\psi}{\partial\theta} = 0, \tag{12.3-5}$$

$$\text{at } r = R \qquad \frac{\partial\psi}{\partial r} = 0. \tag{12.3-6}$$

The velocity of a fluid approaching the sphere is v_∞, i.e.,

$$\text{as } r \to \infty \qquad v_z = v_\infty. \tag{12.3-7}$$

The boundary condition expressed by Eq. (12.3-7) can be expressed in terms of the stream function as follows. From Figure 12.2, the velocity components v_r and v_θ are expressed in terms of v_∞ as

$$v_r = v_\infty\cos\theta = -\frac{1}{r^2\sin\theta}\frac{\partial\psi}{\partial\theta}, \tag{12.3-8}$$

$$v_\theta = -v_\infty\sin\theta = \frac{1}{r\sin\theta}\frac{\partial\psi}{\partial r}. \tag{12.3-9}$$

Integration of Eq. (12.3-8) gives

$$\psi = -\frac{1}{2}v_\infty r^2\sin^2\theta + K(r). \tag{12.3-10}$$

Substitution of Eq. (12.3-10) into Eq. (12.3-9) yields

$$\frac{dK}{dr} = 0 \qquad \Rightarrow \qquad K = \text{constant}. \tag{12.3-11}$$

Since the stream function ψ is arbitrary to a constant, taking $K = 0$ gives

$$\psi = -\frac{1}{2}v_\infty r^2\sin^2\theta \qquad \text{as } r \to \infty. \tag{12.3-12}$$

The boundary condition given by Eq. (12.3-12) suggests that the stream function might be of the form

$$\psi = u(r) \sin^2 \theta. \tag{12.3-13}$$

Substitution of Eq. (12.3-13) into Eq. (12.3-2) yields the following linear, homogeneous, fourth-order differential equation:

$$\frac{d^4 u}{dr^4} - \frac{4}{r^2} \frac{d^2 u}{dr^2} + \frac{8}{r^3} \frac{du}{dr} - \frac{8u}{r^4} = 0. \tag{12.3-14}$$

Equation (12.3-14) is a special form of Euler equation and one form of the solution is given by

$$u = C r^n. \tag{12.3-15}$$

Substitution of Eq. (12.3-15) into Eq. (12.3-14) gives

$$n^4 - 6n^3 + 7n^2 + 6n - 8 = 0. \tag{12.3-16}$$

The roots of Eq. (12.3-16) are

$$n_1 = -1 \qquad n_2 = 1 \qquad n_3 = 2 \qquad n_4 = 4, \tag{12.3-17}$$

so that the solution becomes

$$u(r) = \frac{A}{r} + Br + Cr^2 + Dr^4. \tag{12.3-18}$$

Substitution of Eq. (12.3-18) into Eq. (12.3-13) leads to

$$\psi = \left(\frac{A}{r^3} + \frac{B}{r} + C + Dr^2 \right) r^2 \sin^2 \theta. \tag{12.3-19}$$

Application of the boundary condition expressed by Eq. (12.3-12) gives

$$C = -\frac{1}{2} v_\infty \qquad \text{and} \qquad D = 0. \tag{12.3-20}$$

Thus, Eq. (12.3-19) becomes

$$\psi = \left(\frac{A}{r} + Br - \frac{1}{2} v_\infty r^2 \right) \sin^2 \theta. \tag{12.3-21}$$

Using the boundary conditions at $r = R$ given by Eqs. (12.3-5) and (12.3-6), the constants A and B are evaluated as

$$A = -\frac{1}{4} v_\infty R^3 \qquad \text{and} \qquad B = \frac{3}{4} v_\infty R. \tag{12.3-22}$$

Therefore, the stream function is expressed in the following form:

$$\psi = v_\infty R^2 \sin^2 \theta \left[-\frac{1}{4} \left(\frac{R}{r} \right) + \frac{3}{4} \left(\frac{r}{R} \right) - \frac{1}{2} \left(\frac{r}{R} \right)^2 \right]. \qquad (12.3\text{-}23)$$

The velocity components are then obtained from Eqs. (12.3-3) and (12.3-4) as

$$\frac{v_r}{v_\infty} = \left[1 - \frac{3}{2} \left(\frac{R}{r} \right) + \frac{1}{2} \left(\frac{R}{r} \right)^3 \right] \cos \theta, \qquad (12.3\text{-}24)$$

$$\frac{v_\theta}{v_\infty} = - \left[1 - \frac{3}{4} \left(\frac{R}{r} \right) - \frac{1}{4} \left(\frac{R}{r} \right)^3 \right] \sin \theta. \qquad (12.3\text{-}25)$$

12.3.1　Pressure distribution

To determine the pressure distribution, we have to write the r- and θ-components of the equation of motion. Equations (D) and (E) in Table 1.7 simplify to

$$\frac{\partial P}{\partial r} = \mu \left[\frac{1}{r^2} \frac{\partial}{\partial r} \left(r^2 \frac{\partial v_r}{\partial r} \right) + \frac{1}{r^2 \sin \theta} \frac{\partial}{\partial \theta} \left(\sin \theta \frac{\partial v_r}{\partial \theta} \right) - \frac{2}{r^2} v_r \right.$$

$$\left. - \frac{2}{r^2} \frac{\partial v_\theta}{\partial \theta} - \frac{2}{r^2} v_\theta \cot \theta \right] - \rho g \cos \theta, \qquad (12.3\text{-}26)$$

$$\frac{1}{r} \frac{\partial P}{\partial \theta} = \mu \left[\frac{1}{r^2} \frac{\partial}{\partial r} \left(r^2 \frac{\partial v_\theta}{\partial r} \right) + \frac{1}{r^2 \sin \theta} \frac{\partial}{\partial \theta} \left(\sin \theta \frac{\partial v_\theta}{\partial \theta} \right) + \frac{2}{r^2} \frac{\partial v_r}{\partial \theta} \right.$$

$$\left. - \frac{v_\theta}{r^2 \sin^2 \theta} \right] + \rho g \sin \theta. \qquad (12.3\text{-}27)$$

Substitution of Eqs. (12.3-24) and (12.3-25) into Eqs. (12.3-26) and (12.3-27) gives

$$\frac{\partial P}{\partial r} = 3 \mu v_\infty \cos \theta \frac{R}{r^3} - \rho g \cos \theta, \qquad (12.3\text{-}28)$$

$$\frac{\partial P}{\partial \theta} = \frac{3}{2} \mu v_\infty \sin \theta \frac{R}{r^2} + \rho g r \sin \theta. \qquad (12.3\text{-}29)$$

Integration of Eq. (12.3-28) results in

$$P = -\frac{3}{2} \mu v_\infty \cos \theta \frac{R}{r^2} - \rho g r \cos \theta + f(\theta). \qquad (12.3\text{-}30)$$

The use of Eq. (12.3-30) in Eq. (12.3-29) yields

$$\frac{df}{d\theta} = 0 \quad \Rightarrow \quad f = \text{constant} = C_1. \tag{12.3-31}$$

Therefore, the pressure distribution takes the following form:

$$P = -\frac{3}{2}\mu v_\infty \cos\theta \, \frac{R}{r^2} - \rho g r \cos\theta + C_1. \tag{12.3-32}$$

Application of the boundary condition

$$\text{at } r = \infty \,\&\, \theta = \frac{\pi}{2} \qquad P = P_o, \tag{12.3-33}$$

gives the pressure distribution as

$$P = P_o - \frac{3}{2}\mu v_\infty \cos\theta \, \frac{R}{r^2} - \rho g r \cos\theta. \tag{12.3-34}$$

12.3.2 *Force exerted on the sphere*

From Table 1.4, the non-zero shear stress components are

$$\tau_{rr} = -2\mu \frac{\partial v_r}{\partial r} = -3\mu v_\infty \left[\frac{R}{r^2} - \left(\frac{R}{r}\right)^3 \right] \cos\theta, \tag{12.3-35}$$

$$\tau_{r\theta} = -\mu r \frac{\partial}{\partial r}\left(\frac{v_\theta}{r}\right) = \frac{\mu v_\infty}{r}\left[\frac{3}{2}\frac{R}{r} + \left(\frac{R}{r}\right)^3 - 1 \right] \sin\theta. \tag{12.3-36}$$

The values of these shear stress components on the sphere surface are

$$\tau_{rr}\big|_{r=R} = 0 \qquad \text{and} \qquad \tau_{r\theta}\big|_{r=R} = \frac{3}{2}\frac{\mu v_\infty}{r} \sin\theta. \tag{12.3-37}$$

The details of the calculation of the force exerted by the fluid on the sphere surface were given in Section 8.2.2. Thus, the z-component of the force acting on the sphere is given by

$$F_z = -\underbrace{\int_0^{2\pi}\int_0^{\pi} \cos\theta \, P\big|_{r=R} \, R^2 \sin\theta \, d\theta \, d\phi}_{\text{I}}$$

$$+ \underbrace{\int_0^{2\pi}\int_0^{\pi} \sin\theta \, \mathbf{e}_r\mathbf{e}_\theta{:}\boldsymbol{\tau}\big|_{r=R} \, R^2 \sin\theta \, d\theta \, d\phi}_{\text{II}}. \tag{12.3-38}$$

- **Evaluation of the term I**
  Substitution of Eq. (12.3-34) into Eq. (12.3-38) gives

$$I = -2\pi R^2 P_o \underbrace{\int_0^\pi \cos\theta \sin\theta \, d\theta}_{0}$$

$$+ 2\pi R^2 \left(\frac{3}{2}\frac{\mu v_\infty}{R} + \rho g R\right) \underbrace{\int_0^\pi \cos^2\theta \sin\theta \, d\theta}_{2/3}. \qquad (12.3\text{-}39)$$

Thus,

$$I = 2\pi \mu v_\infty R + \frac{4}{3}\pi R^3 \rho g. \qquad (12.3\text{-}40)$$

The first and second terms on the right-hand side of Eq. (12.3-40) represent form drag and buoyant force, respectively.

- **Evaluation of the term II**
  Substitution of Eq. (12.3-36) into Eq. (12.3-38) gives

$$II = 3\pi \mu v_\infty R \underbrace{\int_0^\pi \sin^3\theta \, d\theta}_{4/3} = 4\pi\pi \mu v_\infty R, \qquad (12.3\text{-}41)$$

which represents friction drag.

Summation of Eqs. (12.3-40) and (12.3-41) leads to

$$F_z = \underbrace{6\pi \mu v_\infty R}_{\text{Drag force}} + \underbrace{\frac{4}{3}\pi R^3 \rho g}_{\text{Buoyancy force}}. \qquad (12.3\text{-}42)$$

In the literature, $F_D = 6\pi \mu v_\infty R$ is known as *Stokes' law* for the force of a fluid on a sphere in the creeping flow region, for which

$$Re = \frac{2R v_\infty \mu}{\rho} < 0.1. \qquad (12.3\text{-}43)$$

The friction factor is expressed as

$$f = \frac{F_D}{A_{ch} K_{ch}} = \frac{6\pi \mu v_\infty R}{(\pi R^2)\left(\frac{1}{2}\rho v_\infty^2\right)} = \frac{24\mu}{D v_\infty \rho} = \frac{24}{Re}. \qquad (12.3\text{-}44)$$

Chapter 13

Macroscopic Balances

Integrations of the equation of continuity, equation of motion, and mechanical energy equation over an arbitrary engineering volume exchanging mass and energy with the surroundings lead to mass balance, momentum balance, and mechanical energy balance at the macroscopic level. Integration of microscopic level equations over the volume of the system eliminates position dependence and the resulting macroscopic level equations appear as ordinary differential equations, with time as the only independent variable.

13.1 Pseudo-Steady-State Approximation

In the application of macroscopic equations, the neglect of the unsteady-state term is often referred to as the *pseudo-steady-state* (or *quasi-steady-state*) approximation. Although this assumption simplifies the mathematical treatment, it cannot be used arbitrarily. Pseudo-steady-state approximation is valid only if the constraint developed in the following analysis is satisfied.

The Navier–Stokes equation is given by Eq. (1.2-22), i.e.,

$$\rho\left(\frac{\partial \mathbf{v}}{\partial t} + \mathbf{v}\cdot\nabla\mathbf{v}\right) = -\nabla P + \mu\nabla^2\mathbf{v} + \rho\mathbf{g}. \tag{13.1-1}$$

Among the terms appearing on the right-hand side of Eq. (13.1-1), the viscous term is the slowest one, with the following order of magnitude:

$$O\left(\mu\nabla^2\mathbf{v}\right) \sim \frac{\mu v_{\text{ch}}}{L_{\text{ch}}^2}. \tag{13.1-2}$$

On the other hand, the order of magnitude of the unsteady-steady (or accumulation) term is

$$O\left(\rho \frac{\partial \mathbf{v}}{\partial t}\right) \sim \frac{\rho v_{\mathrm{ch}}}{t_{\mathrm{ch}}}. \tag{13.1-3}$$

The unsteady-state term becomes negligible when

$$\frac{\rho v_{\mathrm{ch}}}{t_{\mathrm{ch}}} \ll \frac{\mu v_{\mathrm{ch}}}{L_{\mathrm{ch}}^2} \quad \Rightarrow \quad \frac{\mu t_{\mathrm{ch}}}{\rho L_{\mathrm{ch}}^2} \gg 1. \tag{13.1-4}$$

13.2   Macroscopic Mass Balance

The macroscopic mass balance is obtained by integrating the equation of continuity, Eq. (1.2-1), over the volume of the system, $V(t)$. The result is

$$\underbrace{\frac{d}{dt} \int_{V(t)} \rho \, dV}_{\mathrm{I}} + \underbrace{\int_{A_e(t)} \rho (\mathbf{v} - \mathbf{w}) \cdot \mathbf{n} \, dA}_{\mathrm{II}} = 0,. \tag{13.2-1}$$

where $A_e(t)$ represents the area of entrances and exits through which fluid may enter and/or leave the system volume, $\mathbf{w}$ is the velocity of the surface enclosing $V(t)$, and $\mathbf{n}$ is the outwardly directed unit normal vector to $A_e(t)$. Term I is the time rate of change of mass in $V(t)$ and term II is the net rate of mass leaving $V(t)$ through $A_e(t)$. Under steady conditions, term I becomes zero.

13.2.1   *Sudden contraction in a pipeline*

An incompressible liquid is flowing under steady conditions from a large circular pipe into a smaller one as shown in Figure 13.1. If the average velocity at the inlet is 5 cm/s, calculate the average velocity at the exit.

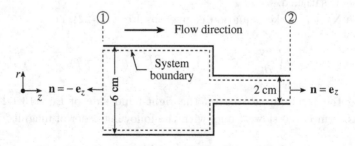

Figure 13.1   Sudden contraction in a pipeline.

Since the system boundary is fixed, then $\mathbf{w} = 0$ and Eq. (13.2-1) simplifies to

$$\rho \int_{A_1} \mathbf{v} \cdot \mathbf{n} \, dA + \rho \int_{A_2} \mathbf{v} \cdot \mathbf{n} \, dA = 0. \qquad (13.2\text{-}2)$$

Noting that

$$\mathbf{v} \cdot \mathbf{n} = \begin{cases} v_z \mathbf{e}_z \cdot (-\mathbf{e}_z) = -v_z & \text{over } A_1, \\ v_z \mathbf{e}_z \cdot (\mathbf{e}_z) = v_z & \text{over } A_2, \end{cases} \qquad (13.2\text{-}3)$$

Eq. (13.2-2) becomes

$$-\int_{A_1} v_z \, dA + \int_{A_2} v_z \, dA = 0. \qquad (13.2\text{-}4)$$

Application of the mean value theorem in two dimensions[1] yields

$$-\langle v_z \rangle_1 A_1 + \langle v_z \rangle_2 A_2 = 0. \qquad (13.2\text{-}5)$$

Thus,

$$\langle v_z \rangle_2 = \langle v_z \rangle_1 \left(\frac{A_1}{A_2} \right) = \langle v_z \rangle_1 \left(\frac{D_1}{D_2} \right)^2. \qquad (13.2\text{-}6)$$

Substitution of the numerical values gives

$$\langle v_z \rangle_2 = (5) \left(\frac{6}{2} \right)^2 = 45 \text{ cm/s}.$$

Comment: Note that Eq. (13.2-1) is the mathematical statement of the conservation of total mass expressed as

$$\begin{pmatrix} \text{Rate of} \\ \text{mass in} \end{pmatrix} - \begin{pmatrix} \text{Rate of} \\ \text{mass out} \end{pmatrix} = \begin{pmatrix} \text{Rate of mass} \\ \text{accumulation} \end{pmatrix}, \qquad (13.2\text{-}7)$$

in which

$$\begin{pmatrix} \text{Rate of} \\ \text{mass in/out} \end{pmatrix} = (\text{Density})(\text{Average velocity})(\text{Area}) = \rho \langle v \rangle A \qquad (13.2\text{-}8)$$

and

$$\begin{pmatrix} \text{Rate of mass} \\ \text{accumulation} \end{pmatrix} = \begin{pmatrix} \text{Time rate of change of} \\ \text{total mass in the system} \end{pmatrix} = \frac{d(\rho V)}{dt}. \qquad (13.2\text{-}9)$$

---

[1] The mean value theorem in two dimensions states that

$$\langle f \rangle = \frac{1}{A} \int_A f \, dA.$$

13.2.2 *Flow of a liquid from one tank to another*

Two identical vertical tanks placed on a platform are connected at the bottom by a horizontal pipe 5 cm in diameter as shown in Figure 13.2. Each tank is open to the atmosphere and 1.5 m in height and 80 cm in diameter. Initially the valve is closed and Tank 1 is full while Tank 2 is empty. When the valve is opened, the average velocity through the pipe is given by

$$\langle v \rangle = 1.8\sqrt{h}, \tag{13.2-10}$$

where $\langle v \rangle$ is the average velocity in m/s and h is the difference between the levels in the two tanks in meters. Calculate the time for the levels in the two tanks to become equal.

Choosing Tank 1 as a system, the conservation of mass is expressed as

$$-\text{Rate of mass out} = \text{Rate of mass accumulation}, \tag{13.2-11}$$

or

$$-\rho(1.8\sqrt{h})\left(\frac{\pi D_o^2}{4}\right) = \frac{d}{dt}\left[\rho(h+H)\left(\frac{\pi D_T^2}{4}\right)\right], \tag{13.2-12}$$

where D_o is the diameter of the pipe connecting the tanks and D_T is the diameter of the tank. Simplification of Eq. (13.2-12) gives

$$-1.8\sqrt{h}\left(\frac{D_o}{D_T}\right)^2 = \frac{dh}{dt} + \frac{dH}{dt}. \tag{13.2-13}$$

Choosing Tank 2 as a system, the conservation of mass is expressed as

$$\text{Rate of mass in} = \text{Rate of mass accumulation}, \tag{13.2-14}$$

or

$$\rho(1.8\sqrt{h})\left(\frac{\pi D_o^2}{4}\right) = \frac{d}{dt}\left[\rho H\left(\frac{\pi D_T^2}{4}\right)\right]. \tag{13.2-15}$$

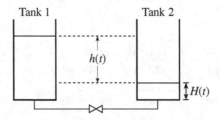

Figure 13.2   Flow of a liquid from one tank to another.

Simplification of Eq. (13.2-15) yields

$$1.8\sqrt{h}\left(\frac{D_o}{D_T}\right)^2 = \frac{dH}{dt}. \tag{13.2-16}$$

Subtraction of Eq. (13.2-16) from Eq. (13.2-13) gives

$$-3.6\sqrt{h}\left(\frac{D_o}{D_T}\right)^2 = \frac{dh}{dt}. \tag{13.2-17}$$

Equation (13.2-17) is a separable equation. Rearrangement results in

$$\int_0^t dt = -\frac{1}{3.6}\left(\frac{D_T}{D_o}\right)^2 \int_{1.5}^0 \frac{dh}{\sqrt{h}}. \tag{13.2-18}$$

Integration and substitution of the numerical values give

$$t = -\frac{1}{1.8}\left(\frac{80}{5}\right)^2 \sqrt{h}\Big|_{1.5}^0 = 174.2 \text{ s} \simeq 3 \text{ min}. \tag{13.2-19}$$

13.2.3 *Removal of liquid from a cylindrical hole*

Consider a piston of diameter D_1 being pushed into a liquid-filled cylindrical hole at a constant velocity U, forcing the incompressible liquid out through a tube of diameter D_2 and length L as shown in Figure 13.3. We want to calculate the average exit velocity of the liquid.

For the system whose boundaries are shown in Figure 13.3, Eq. (13.2-1) reduces to

$$\frac{dV}{dt} + \int_{A_2} \mathbf{v}\cdot\mathbf{n}\,dA = 0. \tag{13.2-20}$$

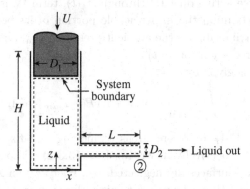

Figure 13.3   Liquid removal from a cylindrical hole.

At any instant, the volume of the liquid within the system is

$$V = \left(\frac{\pi D_1^2}{4}\right) H + \left(\frac{\pi D_2^2}{4}\right) L - \left(\frac{\pi D_1^2}{4}\right) Ut. \tag{13.2-21}$$

Over A_2

$$\mathbf{v} \cdot \mathbf{n} = v_2 \mathbf{e}_x \cdot \mathbf{e}_x = v_2. \tag{13.2-22}$$

Substitution of Eqs. (13.2-21) and (13.2-22) into Eq. (13.2-20) and application of the mean value theorem in two dimensions yield

$$-\left(\frac{\pi D_1^2}{4}\right) U + \langle v_2 \rangle \left(\frac{\pi D_2^2}{4}\right) = 0 \quad \Rightarrow \quad \langle v_2 \rangle = \left(\frac{D_1}{D_2}\right)^2 U. \tag{13.2-23}$$

13.3 Macroscopic Momentum Balance

The macroscopic momentum balance is obtained by integrating the equation of motion, Eq. (1.2-22), over the volume of the system, $V(t)$. The result is

$$\underbrace{\frac{d}{dt} \int_{V(t)} \rho \mathbf{v}\, dV}_{\text{I}} + \underbrace{\int_{A_e(t)} \rho \mathbf{v}(\mathbf{v} - \mathbf{w}) \cdot \mathbf{n}\, dA}_{\text{II}}$$

$$= \underbrace{-\int_{A_e(t)} \mathbf{n}\,(P - P_{\text{atm}})\, dA - \underbrace{\mathbf{F}}_{\text{IV}}}_{\text{III}} + \underbrace{\int_{V(t)} \rho \mathbf{g}\, dV}_{\text{V}}, \tag{13.3-1}$$

where term I is the time rate of change of momentum in $V(t)$, term II is the net rate of momentum leaving $V(t)$ through $A_e(t)$, term III is the net pressure force acting on $V(t)$ through $A_e(t)$, term IV is the force that the system exerts upon the impermeable portion of its bounding surface beyond force attributable to the ambient pressure P_{atm}, and term V is the gravitational force acting on $V(t)$.

The term $\mathbf{F}$ is given by

$$\mathbf{F} = \int_{A_m(t)+A_f} \left[\mathbf{n} \cdot \boldsymbol{\tau} + \mathbf{n}\,(P - P_{\text{atm}}) \right] dA, \tag{13.3-2}$$

where $A_m(t)$ and A_f represent areas of moving and fixed surfaces of a system volume, respectively. In writing Eq. (13.3-2), viscous forces on the entrance and exit surfaces are neglected, i.e., $\mathbf{n} \cdot \boldsymbol{\tau} \approx 0$ on $A_e(t)$. This can be achieved by selecting entrance and exit surfaces as cross-sections normal to flow in long straight pipes.

13.3.1 *Force exerted by a jet on a stationary flat plate*

Consider a jet of an incompressible liquid impinging on a stationary flat plate of height L and width W as shown in Figure 13.4. We want to calculate the force produced by a jet on a flat plate under steady conditions.

Once the liquid jet strikes the plate, it is deflected at right angles. Since pressure in a free jet is atmospheric, Eq. (13.3-1) simplifies to

$$\int_{A_1} \rho \mathbf{v}(\mathbf{v} \cdot \mathbf{n})\, dA + \int_{A_2} \rho \mathbf{v}(\mathbf{v} \cdot \mathbf{n})\, dA + \int_{A_3} \rho \mathbf{v}(\mathbf{v} \cdot \mathbf{n})\, dA$$
$$= -\mathbf{F} + \int_V \rho \mathbf{g}\, dV. \tag{13.3-3}$$

Noting that $\mathbf{v}$ and $\mathbf{n}$ are in opposite directions on A_1 and in the same direction on A_2 and A_3, Eq. (13.3-3) becomes

$$-\int_{A_1} \rho \mathbf{v} v_1\, dA + \int_{A_2} \rho \mathbf{v} v_2\, dA + \int_{A_3} \rho \mathbf{v} v_3\, dA = -\mathbf{F} + \int_V \rho \mathbf{g}\, dV. \tag{13.3-4}$$

Taking the scalar (or dot) product of Eq. (13.3-3) with $\mathbf{e}_x$ gives the x-component of the force as

$$-\int_{A_1} \rho \underbrace{(\mathbf{e}_x \cdot \mathbf{v})}_{v_1} v_1\, dA + \int_{A_2} \rho \underbrace{(\mathbf{e}_x \cdot \mathbf{v})}_{0} v_2\, dA + \int_{A_3} \rho \underbrace{(\mathbf{e}_x \cdot \mathbf{v})}_{0} v_3\, dA$$
$$= -F_x + \int_V \rho \underbrace{(\mathbf{e}_x \cdot \mathbf{g})}_{0}\, dV. \tag{13.3-5}$$

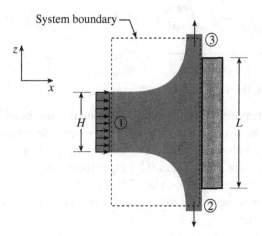

Figure 13.4 Liquid jet impinging on a flat plate.

Therefore,

$$F_x = \rho v_1^2 A_1 = \rho v_1^2 HW. \tag{13.3-6}$$

Numerical Example   A horizontal jet of water 3 cm in diameter strikes a vertical plate. Determine the velocity of the jet if a force of 650 N is required to:

(a) Hold the plate stationary,
(b) Move the plate at a constant velocity of 5 m/s.

Solution

(a) From Eq. (13.3-6),

$$F_x = \rho v_1^2 A_1 = \rho v_1^2 \left(\frac{\pi D_1^2}{4} \right).$$

Therefore,

$$v_1 = \sqrt{\frac{4F_x}{\pi \rho D_1^2}} = \sqrt{\frac{(4)(650)}{\pi(1000)(3 \times 10^{-2})^2}} = 30.3 \text{ m/s}.$$

(b) In this case,

$$v_1 - 5 = \sqrt{\frac{4F_x}{\pi \rho D_1^2}} = 30.3 \quad \Rightarrow \quad v_1 = 35.3 \text{ m/s}.$$

13.3.2   *Flow in a U-bend*

Water flows through a horizontal U-shaped pipe bend as shown in Figure 13.5. The average velocity is 18 m/s and the cross-sectional area is constant at a value of 75 cm². The gauge pressures at the entrance and exit of the bend are 0.7 bar and 0.5 bar, respectively. Let us calculate the net horizontal force exerted on the U-bend.

Since the cross-sectional area is constant, i.e., $A_1 = A_2 = A$, under steady conditions

$$\langle v_1 \rangle = \langle v_2 \rangle = v. \tag{13.3-7}$$

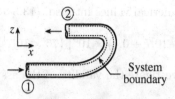

Figure 13.5 Flow in a U-bend.

For the system whose boundaries are shown in Figure 13.5, Eq. (13.3-1) simplifies to

$$\int_{A_1} \rho \mathbf{v}(\mathbf{v} \cdot \mathbf{n}) \, dA + \int_{A_2} \rho \mathbf{v}(\mathbf{v} \cdot \mathbf{n}) \, dA$$

$$= -\int_{A_1} \mathbf{n}\,(P - P_{\text{atm}}) \, dA - \int_{A_2} \mathbf{n}\,(P - P_{\text{atm}}) \, dA - \mathbf{F} + \int_V \rho \mathbf{g} \, dV.$$

$$(13.3\text{-}8)$$

Noting that $\mathbf{v}$ and $\mathbf{n}$ are in opposite directions on A_1 and in the same direction on A_2, Eq. (13.3-8) becomes

$$-\int_{A_1} \rho \mathbf{v} v_1 \, dA + \int_{A_2} \rho \mathbf{v} v_2 \, dA$$

$$= -\int_{A_1} \mathbf{n}\,(P - P_{\text{atm}}) \, dA - \int_{A_2} \mathbf{n}\,(P - P_{\text{atm}}) \, dA - \mathbf{F} + \int_V \rho \mathbf{g} \, dV.$$

$$(13.3\text{-}9)$$

The scalar product of Eq. (13.3-9) with $\mathbf{e}_x$ gives the x-component of the force, F_x. Noting that

$$\text{on } A_1 = \begin{cases} \mathbf{e}_x \cdot \mathbf{v} = v_1 \\ \mathbf{e}_x \cdot \mathbf{n} = -1 \end{cases} \quad \text{and} \quad \text{on } A_2 = \begin{cases} \mathbf{e}_x \cdot \mathbf{v} = -v_2 \\ \mathbf{e}_x \cdot \mathbf{n} = -1 \end{cases}, \quad (13.3\text{-}10)$$

Eq. (13.3-9) takes the following form:

$$-\rho \langle v_1^2 \rangle A - \rho \langle v_2^2 \rangle A = (P_1 - P_{\text{atm}})A + (P_2 - P_{\text{atm}})A - F_x. \quad (13.3\text{-}11)$$

Assuming

$$\langle v_1^2 \rangle \simeq \langle v_1 \rangle^2 \quad \text{and} \quad \langle v_2^2 \rangle \simeq \langle v_2 \rangle^2, \quad (13.3\text{-}12)$$

and making use of Eq. (13.3-7) give the net force as

$$F_x = \left[2\rho v^2 + (P_1 - P_{\text{atm}}) + (P_2 - P_{\text{atm}}) \right] A. \quad (13.3\text{-}13)$$

Substitution of the numerical values into Eq. (13.3-13) yields

$$F_x = \left[2(1000)(18)^2 + (0.7 + 0.5) \times 10^5\right] (75 \times 10^{-4}) = 5760 \text{ N}. \quad (13.3\text{-}14)$$

Comment: Let us investigate the validity of the assumption expressed by Eq. (13.3-12).

For a steady laminar flow of an incompressible fluid in a circular pipe of radius R, the velocity distribution is given by Eq. (5.2-31), i.e.,

$$v_z = \frac{(\mathcal{P}_o - \mathcal{P}_L)R^2}{4\mu L} \left[1 - \left(\frac{r}{R}\right)^2\right]. \quad (13.3\text{-}15)$$

Since the maximum velocity, $v_{\max}$, occurs at the pipe center, Eq. (13.3-15) becomes

$$v_z = v_{\max}\left[1 - \left(\frac{r}{R}\right)^2\right], \quad (13.3\text{-}16)$$

where

$$v_{\max} = \frac{(\mathcal{P}_o - \mathcal{P}_L)R^2}{4\mu L}. \quad (13.3\text{-}17)$$

The average velocity is given by

$$\langle v_z \rangle = \frac{1}{\pi R^2} \int_0^{2\pi} \int_0^R v_z r\, dr\, d\theta. \quad (13.3\text{-}18)$$

Substitution of Eq. (13.3-16) into Eq. (13.3-18) and integration lead to

$$\langle v_z \rangle = \frac{v_{\max}}{2}. \quad (13.3\text{-}19)$$

On the other hand, the average of the square of the velocity is given by

$$\langle v_z^2 \rangle = \frac{1}{\pi R^2} \int_0^{2\pi} \int_0^R v_z^2 r\, dr\, d\theta. \quad (13.3\text{-}20)$$

Substitution of Eq. (13.3-16) into Eq. (13.3-20) and integration lead to

$$\langle v_z^2 \rangle = \frac{v_{\max}^2}{3}. \quad (13.3\text{-}21)$$

Thus,

$$\frac{\langle v_z^2 \rangle}{\langle v_z \rangle^2} = \frac{4}{3} = 1.333. \quad (13.3\text{-}22)$$

For steady turbulent flow of an incompressible fluid in a circular pipe of radius R, the time-averaged velocity distribution is expressed as

$$\overline{v}_z = \overline{v}_{\max} \left(1 - \frac{r}{R}\right)^{1/n}, \tag{13.3-23}$$

where the overbar indicates time-averaged velocities and the exponent n is dependent on the Reynolds number. In practice, $n = 7$ approximates many flows. The average velocity is given by

$$\langle \overline{v}_z \rangle = \frac{1}{\pi R^2} \int_0^{2\pi} \int_0^R \overline{v}_z r \, dr \, d\theta. \tag{13.3-24}$$

Substitution of Eq. (13.3-23) into Eq. (13.3-24) and integration give

$$\langle \overline{v}_z \rangle = \frac{2n^2}{(n+1)(2n+1)} v_{\max}. \tag{13.3-25}$$

On the other hand, the average of the square of the velocity is given by

$$\langle \overline{v}_z^2 \rangle = \frac{1}{\pi R^2} \int_0^{2\pi} \int_0^R \overline{v}_z^2 r \, dr \, d\theta. \tag{13.3-26}$$

Substitution of Eq. (13.3-23) into Eq. (13.3-26) and integration lead to

$$\langle \overline{v}_z^2 \rangle = \frac{n^2}{(n+1)(n+2)} v_{\max}^2. \tag{13.3-27}$$

Hence,

$$\frac{\langle \overline{v}_z^2 \rangle}{\langle \overline{v}_z \rangle^2} = \frac{(n+1)(2n+1)^2}{4n^2(n+2)}. \tag{13.3-28}$$

For $n = 7$, Eq. (13.3-28) gives

$$\frac{\langle \overline{v}_z^2 \rangle}{\langle \overline{v}_z \rangle^2} = 1.020. \tag{13.3-29}$$

Therefore, while the assumption expressed by Eq. (13.3-12) is valid for turbulent flow, it leads to an error for laminar flow.

The diameter of the U-bend is

$$D = \sqrt{\frac{(4)(75 \times 10^{-4})}{\pi}} = 0.098 \text{ m}.$$

The Reynolds number is

$$\text{Re} = \frac{(0.098)(18)(1000)}{1000 \times 10^{-6}} = 1{,}764{,}000.$$

Therefore, the flow is turbulent.

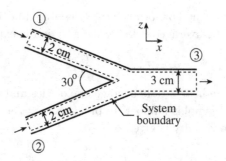

Figure 13.6   Mixing of liquids in a Y-junction.

13.3.3   *Mixing of liquids in a Y-junction*

Consider the mixing of two water streams in a Y-junction as shown in Figure 13.6. Using the following given data, let us calculate the net horizontal and vertical forces exerted by the fluid on the Y-junction:

$$\langle v_1 \rangle = 4.5 \text{ m/s} \quad \text{and} \quad \langle v_2 \rangle = 6 \text{ m/s},$$

$$P_1 - P_{\text{atm}} = 1.1 \text{ bar} \qquad P_2 - P_{\text{atm}} = 1.5 \text{ bar} \qquad P_3 - P_{\text{atm}} = 0.$$

Under steady conditions, the conservation of mass states that

$$\text{Rate of mass in} = \text{Rate of mass out}, \qquad (13.3\text{-}30)$$

or

$$\rho \langle v_1 \rangle \left(\frac{\pi D_1^2}{4} \right) + \rho \langle v_2 \rangle \left(\frac{\pi D_2^2}{4} \right) = \rho \langle v_3 \rangle \left(\frac{\pi D_3^2}{4} \right). \qquad (13.3\text{-}31)$$

Substitution of the numerical values into Eq. (13.3-31) gives

$$\langle v_3 \rangle = \frac{(4.5)(2)^2 + (6)(2)^2}{3^2} = 4.67 \text{ m/s}. \qquad (13.3\text{-}32)$$

For the system whose boundaries are shown in Figure 13.6, Eq. (13.3-1) simplifies to

$$\int_{A_1} \rho \mathbf{v}(\mathbf{v} \cdot \mathbf{n}) \, dA + \int_{A_2} \rho \mathbf{v}(\mathbf{v} \cdot \mathbf{n}) \, dA + \int_{A_3} \rho \mathbf{v}(\mathbf{v} \cdot \mathbf{n}) \, dA$$

$$= -\int_{A_1} \mathbf{n} \, (P - P_{\text{atm}}) \, dA - \int_{A_2} \mathbf{n} \, (P - P_{\text{atm}}) \, dA - \int_{A_3} \mathbf{n} \, (P - P_{\text{atm}}) \, dA$$

$$- \mathbf{F} + \int_V \rho \mathbf{g} \, dV. \qquad (13.3\text{-}33)$$

Noting that $\mathbf{v}$ and $\mathbf{n}$ are in opposite directions on A_1 and A_2 and in the same direction on A_3, Eq. (13.3-33) becomes

$$-\int_{A_1} \rho v v_1 \, dA - \int_{A_2} \rho v v_2 \, dA + \int_{A_3} \rho v v_3 \, dA$$

$$= -\int_{A_1} \mathbf{n}\left(P - P_{\text{atm}}\right) dA - \int_{A_2} \mathbf{n}\left(P - P_{\text{atm}}\right) dA$$

$$- \int_{A_3} \mathbf{n}\left(P - P_{\text{atm}}\right) dA - \mathbf{F} + \int_V \rho \mathbf{g} \, dV. \tag{13.3-34}$$

The scalar product of Eq. (13.3-34) with $\mathbf{e}_x$ gives the x-component of the force, F_x. Noting that

$$\text{on } A_1 = \begin{cases} \mathbf{e}_x \cdot \mathbf{v} = v_1 \cos 15 \\ \mathbf{e}_x \cdot \mathbf{n} = \cos\left(\pi - 15\right) = -\cos 15, \end{cases} \tag{13.3-35}$$

$$\text{on } A_2 = \begin{cases} \mathbf{e}_x \cdot \mathbf{v} = v_2 \cos 15 \\ \mathbf{e}_x \cdot \mathbf{n} = \cos(\pi - 15) = -\cos 15, \end{cases} \tag{13.3-36}$$

$$\text{on } A_3 = \begin{cases} \mathbf{e}_x \cdot \mathbf{v} = v_3 \\ \mathbf{e}_x \cdot \mathbf{n} = \cos 0 = 1. \end{cases} \tag{13.3-37}$$

Eq. (13.3-34) reduces to

$$-\rho \cos 15 \int_{A_1} v_1^2 \, dA - \rho \cos 15 \int_{A_2} v_2^2 \, dA + \rho \int_{A_3} v_3^2 \, dA$$

$$= \cos 15(P_1 - P_{\text{atm}})A_1 + \cos 15(P_2 - P_{\text{atm}})A_2 - (P_3 - P_{\text{atm}})A_3 - F_x. \tag{13.3-38}$$

Rearrangement gives

$$F_x = (\cos 15)(P_1 - P_{\text{atm}})A_1 + (\cos 15)(P_2 - P_{\text{atm}})A_2 - (P_3 - P_{\text{atm}})A_3$$

$$+ \rho(\cos 15)\langle v_1 \rangle^2 A_1 + \rho(\cos 15)\langle v_2 \rangle^2 A_2 - \rho\langle v_3 \rangle^2 A_3. \tag{13.3-39}$$

The cross-sectional areas are

$$A_1 = A_2 = \frac{\pi(0.02)^2}{4} = 3.142 \times 10^{-4} \, \text{m}^2 \quad A_3 = \frac{\pi(0.03)^2}{4} = 7.069 \times 10^{-4} \, \text{m}^2.$$

Substitution of the numerical values into Eq. (13.3-39) yields

$$F_x = (\cos 15)(3.142 \times 10^{-4})(1.1 + 1.5) \times 10^5$$

$$+ (1000)(\cos 15)(3.142 \times 10^{-4})(4.5^2 + 6^2)$$

$$- (1000)(4.67)^2(7.069 \times 10^{-4}) = 80.6 \, \text{N}.$$

The scalar product of Eq. (13.3-34) with $\mathbf{e}_z$ gives the z-component of the force, F_z. Noting that

$$\text{on } A_1 = \begin{cases} \mathbf{e}_z \cdot \mathbf{v} = v_1 \cos\left(\dfrac{\pi}{2} + 15\right) = -\sin 15 \\ \mathbf{e}_z \cdot \mathbf{n} = \cos\left(\pi - 105\right) = -\cos 105, \end{cases} \tag{13.3-40}$$

$$\text{on } A_2 = \begin{cases} \mathbf{e}_z \cdot \mathbf{v} = v_2 \cos\left(\dfrac{\pi}{2} - 15\right) = \sin 15 \\ \mathbf{e}_z \cdot \mathbf{n} = \cos\left(\dfrac{\pi}{2} + 15\right) = -\sin 15, \end{cases} \tag{13.3-41}$$

$$\text{on } A_3 = \begin{cases} \mathbf{e}_z \cdot \mathbf{v} = 0 \\ \mathbf{e}_z \cdot \mathbf{n} = \cos 90 = 0. \end{cases} \tag{13.3-42}$$

Eq. (13.3-34) reduces to

$$\rho(\sin 15) \int_{A_1} v_1^2 \, dA - \rho(\cos 15) \int_{A_2} v_2^2 \, dA$$
$$= (\cos 105)(P_1 - P_{\text{atm}})A_1 + (\sin 15)(P_2 - P_{\text{atm}})A_2 - F_z - \rho g V. \tag{13.3-43}$$

Neglecting the gravitational term and rearranging give

$$F_z = (\cos 105)(P_1 - P_{\text{atm}})A_1 + (\sin 15)(P_2 - P_{\text{atm}})A_2$$
$$- \rho(\sin 15)\langle v_1\rangle^2 A_1 + \rho(\sin 15)\langle v_2\rangle^2 A_2. \tag{13.3-44}$$

Substitution of the numerical values into Eq. (13.3-44) gives

$$F_z = \left[(\cos 105)(1.1) + (\sin 15)(1.5)\right] \times 10^5 (3.142 \times 10^{-4})$$
$$+ (1000)(\sin 15)(6^2 - 4.5^2)(3.142 \times 10^{-4}) = 4.5 \text{ N}.$$

13.4 Macroscopic Mechanical Energy Equation

The macroscopic mechanical energy balance is obtained by integrating the equation of motion, Eq. (1.2-31), over the volume of the system, $V(t)$. The result is

$$\underbrace{\frac{d}{dt} \int_{V(t)} \frac{1}{2}\rho v^2 \, dV}_{\text{I}} + \underbrace{\int_{A_e(t)} \frac{1}{2}\rho v^2 (\mathbf{v} - \mathbf{w}) \cdot \mathbf{n} \, dA}_{\text{II}}$$

$$= \underbrace{-\int_{A_e(t)} (P - P_{\text{atm}})(\mathbf{n} \cdot \mathbf{v}) \, dA}_{\text{III}} - \underbrace{W}_{\text{IV}} - \underbrace{E_v}_{\text{V}} - \underbrace{\int_{A(t)} \rho \hat{\phi} \mathbf{v} \cdot \mathbf{n} \, dA}_{\text{VI}}, \tag{13.4-1}$$

where term I is the time rate of change of kinetic energy of the system; term II is the net rate of kinetic energy leaving the system; term III is the rate of pressure work done on the system at entrances and exits, i.e., flow work; term IV is the rate of work done by the system on the surroundings by solid moving surfaces; term V is the rate of viscous dissipation to internal energy; and term VI is the rate of work done on the control volume by gravity.

For a given problem, when macroscopic mechanical energy and momentum balances give the same result, there is a good chance that the result is satisfactory. However, there may be cases in which they lead to different results. The difference mainly stems from the neglect of viscous effects. In the macroscopic momentum balance, the assumption of "negligible viscous effects" implies $\mathbf{n} \cdot \boldsymbol{\tau} \simeq 0$ over the surface area of the system. In the macroscopic mechanical energy balance, however, the same assumption implies that E_v, defined by

$$E_v = -\int_{V(t)} \boldsymbol{\tau} : \nabla \mathbf{v} \, dV, \qquad (13.4\text{-}2)$$

is negligible. In other words, $\boldsymbol{\tau} : \nabla \mathbf{v} \simeq 0$ over the volume of the system. It is difficult to make an *a priori* decision on which term is negligible.

13.4.1 *Energy generation from a water turbine*

Steady flow of water through the apparatus shown in Figure 13.7 generates energy at a rate of 10 hp. Assuming flat velocity profiles and negligible viscous effects, let us calculate the horizontal force developed on the duct from the flow of water.

Under steady conditions, the conservation of mass states that

$$\begin{pmatrix} \text{Rate of} \\ \text{mass in} \end{pmatrix} = \begin{pmatrix} \text{Rate of} \\ \text{mass out} \end{pmatrix} \quad \Rightarrow \quad \rho \langle v_1 \rangle A_1 = \rho \langle v_2 \rangle A_2, \qquad (13.4\text{-}3)$$

or

$$\langle v_2 \rangle = \langle v_1 \rangle \frac{A_1}{A_2} = \langle v_1 \rangle \left(\frac{D_1}{D_2} \right)^2 = 4.5 \left(\frac{0.6}{0.45} \right)^2 = 8 \text{ m/s.} \qquad (13.4\text{-}4)$$

The macroscopic mechanical energy balance, Eq. (13.4-1), simplifies to

$$\int_{A_1} \frac{1}{2} \rho v^2 (\mathbf{v} \cdot \mathbf{n}) \, dA + \int_{A_2} \frac{1}{2} \rho v^2 (\mathbf{v} \cdot \mathbf{n}) \, dA$$
$$= -\int_{A_1} (P - P_{\text{atm}}) (\mathbf{n} \cdot \mathbf{v}) \, dA - \int_{A_2} (P - P_{\text{atm}}) (\mathbf{n} \cdot \mathbf{v}) \, dA - W.$$

$$(13.4\text{-}5)$$

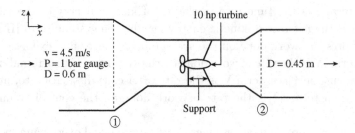

Figure 13.7   Energy generation from a water turbine.

Noting that

$$\mathbf{n} \cdot \mathbf{v} = \begin{cases} -v_1 & \text{on } A_1 \\ v_2 & \text{on } A_2 \end{cases}. \qquad (13.4\text{-}6)$$

Eq. (13.4-5) becomes

$$-\int_{A_1} \frac{1}{2}\rho v^3 \, dA + \int_{A_2} \frac{1}{2}\rho v^3 \, dA = \int_{A_1} (P - P_{\text{atm}})v \, dA$$
$$-\int_{A_2} (P - P_{\text{atm}})v \, dA - W, \qquad (13.4\text{-}7)$$

or[2]

$$-\frac{1}{2}\rho\langle v_1\rangle^3 A_1 + \frac{1}{2}\rho\langle v_2\rangle^3 A_2 = (P_1 - P_{\text{atm}})\langle v_1\rangle A_1$$
$$- (P_2 - P_{\text{atm}})\langle v_2\rangle A_2 - W. \qquad (13.4\text{-}8)$$

Solving for $(P_2 - P_{\text{atm}})$ gives

$$P_2 - P_{\text{atm}} = (P_1 - P_{\text{atm}}) \left(\frac{\langle v_1\rangle A_1}{\langle v_2\rangle A_2} \right) + \frac{1}{2}\rho \left(\frac{\langle v_1\rangle^3 A_1 - \langle v_2\rangle^3 A_2}{\langle v_2\rangle A_2} \right)$$
$$- \frac{W}{\langle v_2\rangle A_2}. \qquad (13.4\text{-}9)$$

Using the relationship given by Eq. (13.4-4), Eq. (13.4-9) simplifies to

$$P_2 - P_{\text{atm}} = (P_1 - P_{\text{atm}}) + \frac{1}{2}\rho \left(\langle v_1\rangle^2 - \langle v_2\rangle^2 \right) - \frac{W}{\langle v_2\rangle A_2}. \qquad (13.4\text{-}10)$$

---

[2] Using the similar analysis presented in Section 13.3.2,

$$\frac{\langle \bar{v}_z^3 \rangle}{\langle \bar{v}_z \rangle^3} = \frac{(n+1)^3(2n+1)^3}{4n^4(n+3)(2n+3)}.$$

For $n = 7$, $\langle \bar{v}_z^3 \rangle / \langle \bar{v}_z \rangle^3 = 1.058$.

Substitution of the numerical values into Eq. (13.4-10) yields

$$P_2 - P_{\text{atm}} = 1 \times 10^5 + \frac{1}{2}(1000)(4.5^2 - 8^2) - \frac{(10)(745.7)}{(8)\left[\pi(0.45)^2/4\right]}$$

$$= 7.226 \times 10^4 \text{ bar.}$$

The macroscopic momentum balance, Eq. (13.3-1), simplifies to

$$\int_{A_1} \rho \mathbf{v}(\mathbf{v} \cdot \mathbf{n})\, dA + \int_{A_2} \rho \mathbf{v}(\mathbf{v} \cdot \mathbf{n})\, dA = - \int_{A_1} \mathbf{n}\,(P - P_{\text{atm}})\, dA$$

$$- \int_{A_2} \mathbf{n}\,(P - P_{\text{atm}})\, dA - \mathbf{F} + \int_V \rho \mathbf{g}\, dV. \qquad (13.4\text{-}11)$$

With the help of Eq. (13.4-6), Eq. (13.4-11) becomes

$$- \int_{A_1} \rho \mathbf{v} v_1\, dA + \int_{A_2} \rho \mathbf{v} v_2\, dA = - \int_{A_1} \mathbf{n}\,(P - P_{\text{atm}})\, dA$$

$$- \int_{A_2} \mathbf{n}\,(P - P_{\text{atm}})\, dA - \mathbf{F} + \int_V \rho \mathbf{g}\, dV. \qquad (13.4\text{-}12)$$

Taking the scalar product of Eq. (13.4-12) with $\mathbf{e}_x$ leads to

$$-\rho \langle v_1 \rangle^2 A_1 + \rho \langle v_2 \rangle^2 A_2 = A_1 - (P_2 - P_{\text{atm}}) A_2 - F_x. \qquad (13.4\text{-}13)$$

Solving for F_x gives

$$F_x = \left[\rho \langle v_1 \rangle^2 + (P_1 - P_{\text{atm}})\right] A_1 - \left[\rho \langle v_2 \rangle^2 + (P_2 - P_{\text{atm}})\right] A_2. \qquad (13.4\text{-}14)$$

Substitution of the numerical values into Eq. (13.4-14) results in

$$F_x = \left[(1000)(4.5)^2 + 1 \times 10^5\right]\left[\frac{\pi(0.6)^2}{4}\right]$$

$$- \left[(1000)(8)^2 + 7.226 \times 10^4\right]\left[\frac{\pi(0.45)^2}{4}\right] = 1.233 \times 10^4 \text{ N.}$$

13.5 The Bernoulli Equation

The starting point for the Bernoulli equation is the Euler equation, Eq. (1.2-18), in which viscous effects are considered negligible. Between any two points 1 and 2 on a streamline, the Bernoulli equation is expressed as

$$\frac{1}{2}\left(v_2^2 - v_1^2\right) + \int_{P_1}^{P_2} \frac{dP}{\rho} + g\,(z_2 - z_1) = 0. \qquad (13.5\text{-}1)$$

For an incompressible fluid, Eq. (13.5-1) reduces to

$$\frac{1}{2}\left(v_2^2 - v_1^2\right) + \frac{P_2 - P_1}{\rho} + g\left(z_2 - z_1\right) = 0. \tag{13.5-2}$$

Despite its simplifying assumptions, the Bernoulli equation has a wide range of applications. It simply states that an increase in fluid velocity results in a decrease in pressure. For example, the wings of an airplane are designed so that the velocity of air at the upper surface is greater than that at the lower surface. This implies that the pressure at the lower surface is higher than the pressure at the upper surface. The resulting pressure difference generates a lift force. Using the same argument, try to answer the following questions:

- Why does the shower curtain always move toward water?
- In a subway station, why is it dangerous to stand near the edge of the platform as the train arrives?

13.5.1   Use of a siphon for draining liquids

Consider a siphon to drain water from a large tank as shown in Figure 13.8. Note that the pressure is atmospheric at the free surface of the water in the tank, point 1, and at the lower end of the siphon, point 2. Application of the Bernoulli equation between points 1 and 2 leads to

$$\frac{1}{2}\left(v_2^2 - v_1^2\right) + g(-h - 0) = 0. \tag{13.5-3}$$

Since $A_1 \gg A_2$, v_1 is very small compared to v_2 and can be neglected. As a result, Eq. (13.5-3) gives the average velocity of water exiting the siphon as

$$v_2 = \sqrt{2gh}. \tag{13.5-4}$$

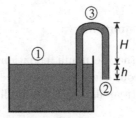

Figure 13.8   Siphoning of water from a large tank.

Therefore, v_2 increases as the tube exit is lowered below the tank surface.

Noting $v_2 = v_3$, application of the Bernoulli equation between points 2 and 3 gives

$$\frac{P_3 - P_{\text{atm}}}{\rho} + g\left[H - (-h)\right] = 0 \quad \Rightarrow \quad P_3 = P_{\text{atm}} - \rho g(h + H). \quad (13.5\text{-}5)$$

Equation (13.5-5) indicates that the pressure at point 3 is less than atmospheric. Since $P_3 > 0$, then

$$P_{\text{atm}} > \rho g(h + H) \quad \Rightarrow \quad h + H < \frac{P_{\text{atm}}}{\rho g} = \frac{1.013 \times 10^5}{(1000)(9.8)} = 10.3 \text{ m}.$$

In other words, the siphon can never lift water more than approximately 10 m. Theoretically, P_3 cannot be equal to zero since water starts to evaporate when $P_3 < P^{\text{vap}}$. The presence of vapor bubbles, known as cavitation, interferes with the flow of water, causing operating problems during the use of a siphon.

13.5.2 *Liquid draining from a vertical cylindrical tank*

Consider a large tank of diameter D filled with liquid as shown in Figure 13.9. A jet of liquid of diameter D_2 flows from a smooth rounded nozzle placed near the bottom of the tank.

- **Macroscopic mass balance**

  Note that $\mathbf{v} = \mathbf{w}$ over A_1 and let us assume that the velocity profile at the exit of the nozzle is flat. Hence, Eq. (13.2-1) simplifies to

  $$\frac{dV}{dt} + \int_{A_2} \mathbf{v} \cdot \mathbf{n} \, dA = 0. \quad (13.5\text{-}6)$$

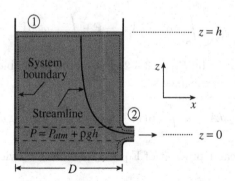

Figure 13.9   Draining of a vertical tank.

Since **v** and **n** are in the same direction on A_2, Eq. (13.5-6) becomes

$$A_1 \frac{dh}{dt} + v_2 A_2 = 0, \tag{13.5-7}$$

where A_1 is the cross-sectional area of the tank. Note that the velocity of the free surface is given by

$$v_1 = -\frac{dh}{dt}. \tag{13.5-8}$$

Thus, Eq. (13.5-7) becomes

$$v_1 A_1 = v_2 A_2 \quad \Rightarrow \quad v_1 = v_2 \left(\frac{A_2}{A_1} \right). \tag{13.5-9}$$

Since $A_2 \ll A_1$, Eq. (13.5-9) implies that $v_1 \ll v_2$. In other words, the height of the liquid in the tank changes very slowly.

- **Macroscopic momentum balance**

Since the liquid height changes very slowly, we will make use of the pseudo-steady-state approximation, i.e.,

$$\frac{d}{dt} \int_{V(t)} \rho \mathbf{v} \, dV \simeq 0. \tag{13.5-10}$$

It is difficult to justify the validity of this assumption at this stage. Once the solution is obtained for the outlet liquid velocity, one should calculate the variation in liquid height in the tank with time and check whether the assumption is reasonable.

Since pressure is atmospheric on A_1 and A_2, Eq. (13.3-1) reduces to

$$\int_{A_2} \rho \mathbf{v} v_2 \, dA = -\mathbf{F} + \int_{V(t)} \rho \mathbf{g} \, dV, \tag{13.5-11}$$

where **F** is given by Eq. (13.3-2). Assuming negligible viscous effects, i.e., $\mathbf{n} \cdot \boldsymbol{\tau} \simeq 0$ over the fixed surfaces of the system, Eq. (13.5-11) becomes

$$\int_{A_2} \rho \mathbf{v} v_2 \, dA = -\int_{A_f} \mathbf{n}(P - P_{\text{atm}}) \, dA + \int_{V(t)} \rho \mathbf{g} \, dV. \tag{13.5-12}$$

Taking the scalar product of Eq. (13.5-12) with $\mathbf{e}_x$ yields

$$\int_{A_2} \rho v_2^2 \, dA = -\int_{A_{\text{right}}} (P - P_{\text{atm}}) \, dA + \int_{A_{\text{left}}} (P - P_{\text{atm}}) \, dA. \tag{13.5-13}$$

The fixed surface on the left-hand side of the tank, A_{left}, is greater than the fixed surface on the right-hand side, A_{right}, by A_2. Thus, Eq. (13.5-13) becomes

$$\int_{A_2} \rho v_2^2 \, dA = \int_{A_2} (P - P_{\text{atm}}) \, dA. \qquad (13.5\text{-}14)$$

If the pressure in the tank is hydrostatic, then

$$P - P_{\text{atm}} = \rho g h \quad \text{on } A_2. \qquad (13.5\text{-}15)$$

Substitution of Eq. (13.5-15) into Eq. (13.5-14) and integration give

$$\rho v_2^2 A_2 = \rho g h A_2 \qquad \Longrightarrow \qquad v_2 = \sqrt{gh}. \qquad (13.5\text{-}16)$$

• Macroscopic mechanical energy balance

Application of the pseudo-steady-state approximation leads to

$$\frac{d}{dt} \int_{V(t)} \frac{1}{2} \rho v^2 \, dV \simeq 0. \qquad (13.5\text{-}17)$$

Note that $W = 0$ and let us neglect viscous effects, i.e., $E_v \simeq 0$. Under these circumstances, Eq. (13.4-1) becomes

$$\int_{A_2} \frac{1}{2} \rho v^2 (\mathbf{v} \cdot \mathbf{n}) \, dA = - \int_{A_1} \rho \widehat{\phi} (\mathbf{v} \cdot \mathbf{n}) \, dA - \int_{A_2} \rho \widehat{\phi} (\mathbf{v} \cdot \mathbf{n}) \, dA, \quad (13.5\text{-}18)$$

where

$$\widehat{\phi} = gz \quad \Rightarrow \quad z = \begin{cases} h & \text{over } A_1 \\ 0 & \text{over } A_2 \end{cases}. \qquad (13.5\text{-}19)$$

While $\mathbf{v}$ and $\mathbf{n}$ are in the same direction over A_2, they are in opposite directions over A_1. Thus, Eq. (13.5-18) becomes

$$\frac{1}{2} \rho v_2^3 A_2 = \rho g h v_1 A_1. \qquad (13.5\text{-}20)$$

Substitution of Eq. (13.5-9) into Eq. (13.5-20) and rearrangement yield

$$v_2 = \sqrt{2gh}. \qquad (13.5\text{-}21)$$

- **Bernoulli equation**

  Application of the Bernoulli equation to a streamline, Eq. (13.5-1), between the free surface (1) and the nozzle (2) gives

  $$\frac{1}{2}\left(v_2^2 - v_1^2\right) + g\left(0 - h\right) = 0. \tag{13.5-22}$$

  From Eq. (13.5-9),

  $$v_2 \gg v_1. \tag{13.5-23}$$

  Thus, $v_2^2 - v_1^2 \simeq v_2^2$ and Eq. (13.5-22) gives

  $$v_2 = \sqrt{2gh}. \tag{13.5-24}$$

  Comment: Experimental results using water indicate that

  $$v_2 = 0.98\sqrt{2gh}. \tag{13.5-25}$$

  Therefore, the results obtained from the macroscopic mechanical energy balance and the Bernoulli equation are very close to the experimental one. On the other hand, the result obtained from the macroscopic momentum balance differs by a factor of $\sqrt{2}$. This may be due to the fact that the pressure in the tank is not hydrostatic everywhere. For a more thorough discussion on this subject, see Whitaker (1992).

  In general, liquid velocity through a nozzle or orifice is related to the liquid height as

  $$v_2 = C\sqrt{2gh}, \tag{13.5-26}$$

  where C is termed *discharge coefficient*.

13.5.3 *Liquid draining from a horizontal cylindrical tank*

A horizontal cylindrical tank of diameter D and length L is filled with liquid as shown in Figure 13.10(a). The liquid leaves the tank through a short circular nozzle of internal diameter D_o in the bottom. How long will it take to drain the tank if the top of the cylinder is open to the atmosphere?

    The conservation statement for mass reduces to

$$- \text{Rate of mass out} = \text{Rate of mass accumulation,} \tag{13.5-27}$$

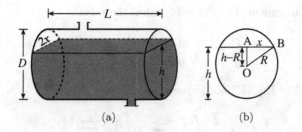

Figure 13.10 Draining of a liquid from a horizontal tank.

or

$$-\rho \langle v_o \rangle A_o = \frac{d(V\rho)}{dt}. \tag{13.5-28}$$

Substitution of Eq. (13.5-26) into Eq. (13.5-28) gives

$$-A_o C \sqrt{2g}\sqrt{h} = \frac{dV}{dt}. \tag{13.5-29}$$

The volume of the liquid in the tank is given by

$$V = \int_0^h 2xL\,dh = 2L \int_0^h x\,dh. \tag{13.5-30}$$

From the triangle OAB in Figure 13.10(b),

$$x^2 = R^2 - (h-R)^2 = 2hR - h^2 = Dh - h^2. \tag{13.5-31}$$

Substitution of Eq. (13.5-31) into Eq. (13.5-30) yields

$$V = 2L \int_0^h \sqrt{Du - u^2}\,du, \tag{13.5-32}$$

where u is a dummy variable of integration. Substitution of Eq. (13.5-32) into Eq. (13.5-29) gives

$$-A_o C \sqrt{2g}\sqrt{h} = 2L \frac{d}{dt}\left(\int_0^h \sqrt{Du - u^2}\,du \right). \tag{13.5-33}$$

Application of the Leibniz rule[3] yields

$$-A_o C \sqrt{2g}\sqrt{h} = 2L\sqrt{Dh - h^2}\,\frac{dh}{dt}. \tag{13.5-34}$$

---

[3]The Leibniz rule for differentiation of integrals is given by

$$\frac{d}{dt}\int_{a(t)}^{b(t)} f(x,t)dx = \int_{a(t)}^{b(t)} \frac{\partial f}{\partial t}dx + f[b(t),t]\frac{db}{dt} - f[a(t),t]\frac{da}{dt}.$$

After rearrangement, Eq. (13.5-34) takes the form

$$\int_0^t dt = \sqrt{\frac{2}{g}} \frac{L}{A_o C} \int_0^h \sqrt{D - h}\, dh. \tag{13.5-35}$$

Integration[4] of Eq. (13.5-35) gives

$$t = \sqrt{\frac{2}{g}} \frac{L}{A_o C} \frac{2}{(-3)} \left. \sqrt{(D - h)^3} \right|_0^h$$

$$= \frac{1}{3} \sqrt{\frac{8}{g}} \frac{L}{A_o C} \left[D^{3/2} - (D - h)^{3/2} \right]. \tag{13.5-36}$$

Since the tank is initially full, i.e., $h = D$, the time efflux is given by

$$t = \frac{1}{3} \sqrt{\frac{8}{g} \frac{L D^{3/2}}{A_o C}}. \tag{13.5-37}$$

Let us calculate the efflux time if $L = 5.5$ m, $D = 1.8$ m, $D_o = 5$ cm, and $C = 1$. The cross-sectional area of the nozzle is

$$A_o = \frac{\pi (5 \times 10^{-2})^2}{4} = 1.963 \times 10^{-3} \text{ m}^2.$$

Substitution of the numerical values into Eq. (13.5-37) yields

$$t = \frac{1}{3} \sqrt{\frac{8}{9.8} \frac{(5.5)(1.8)^{3/2}}{1.963 \times 10^{-3}}} = 2037.8 \text{ s} \simeq 34 \text{ min.}$$

13.5.4 *Liquid draining from a funnel*

A funnel, shown in Figure 13.11, is completely filled with liquid at $t = 0$. The radius of the funnel is given by

$$r = k h^{1/4}, \tag{13.5-38}$$

where k is a known constant. The liquid drains from the funnel through a smooth rounded opening of cross-sectional area A_2, where A_2 is small compared to the inside cross-sectional area of the funnel. Let us assume pseudo-steady-state and neglect viscous effects. Application of the Bernoulli

---

[4]Note that

$$\int \sqrt{ax + b}\, dx = \frac{2\sqrt{(ax + b)^3}}{3a}.$$

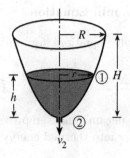

Figure 13.11 Liquid draining from a funnel.

equation to a streamline, Eq. (13.5-1), between the free surface (1) and the outlet (2) gives

$$\frac{1}{2}\left(v_2^2 - v_1^2\right) + g\left(0 - h\right) = 0. \tag{13.5-39}$$

Since $A_1 \gg A_2$, then $v_2 \gg v_1$ and Eq. (13.5-39) leads to

$$v_2 = \sqrt{2gh}. \tag{13.5-40}$$

The conservation statement for mass reduces to

$$- \text{ Rate of mass out} = \text{Rate of mass accumulation}, \tag{13.5-41}$$

or

$$- \rho v_2 A_2 = \frac{d(V\rho)}{dt}. \tag{13.5-42}$$

Substitution of Eq. (13.5-40) into Eq. (13.5-42) and rearrangement yield

$$- \sqrt{2gh}A_2 = \frac{dV}{dt} = \frac{d}{dt}\int_0^h \pi r^2 \, dh. \tag{13.5-43}$$

Application of the Leibniz rule gives

$$- \sqrt{2gh}A_2 = \pi\left[r(h)\right]^2 \frac{dh}{dt} = \pi k^2 \sqrt{h}\,\frac{dh}{dt}. \tag{13.5-44}$$

Simplification of Eq. (13.5-44) and rearrangement result in

$$\int_0^t dt = -\frac{\pi k^2}{A_2\sqrt{2g}}\int_H^0 dh. \tag{13.5-45}$$

Therefore, the time it takes for the funnel to drain completely is given by

$$t = \frac{\pi k^2 H}{A_2\sqrt{2g}}. \tag{13.5-46}$$

13.6   Engineering Bernoulli Equation

The engineering Bernoulli equation for an incompressible fluid is expressed as

$$\frac{\Delta P}{\rho} + \frac{\Delta \langle v \rangle^2}{2} + g\Delta h + \widehat{E}_v - \widehat{W}_s = 0, \qquad (13.6\text{-}1)$$

where $\widehat{E}_v$, the friction loss per unit mass, represents the irreversible degradation of mechanical energy into thermal energy and $\widehat{W}_s$ is the shaft work per unit mass.

For flow in a pipe of diameter D, the friction loss per unit mass is given by

$$\widehat{E}_v = \frac{2 f L \langle v \rangle^2}{D}. \qquad (13.6\text{-}2)$$

The friction factor for laminar flow is given by Eq. (5.2-42). When the flow is turbulent, however, no theoretical solution exists. In this case, the friction factor is usually determined from the *Moody chart* (1944), in which it is expressed as a function of the Reynolds number, Re, and the relative pipe wall roughness, ε/D. Moody prepared this chart by using the equation proposed by Colebrook (1939):

$$\frac{1}{\sqrt{f}} = -4\log\left(\frac{\varepsilon/D}{3.7065} + \frac{1.2613}{\text{Re}\sqrt{f}}\right), \qquad (13.6\text{-}3)$$

where ε is the surface roughness of the pipe wall in meters.

13.6.1   *Draining of a tank with a pipe system*

A cylindrical tank, 5 m in diameter, discharges through a mild steel pipe system ($\varepsilon = 4.6 \times 10^{-5}$ m) connected to the tank base as shown in Figure 13.12.

The drain pipe system has an equivalent length of 80 m and a diameter of 20 cm. Let $h(t)$ be the height of the liquid within the tank with respect to the reference plane and $H = 0.5$ m.

Noting that

$$P_2 = P_1 = P_{\text{atm}} \qquad \text{and} \qquad \langle v_1 \rangle \simeq 0,$$

application of the engineering Bernoulli equation, Eq. (13.6-1), between points 1 and 2 gives

$$\frac{\langle v_2 \rangle^2}{2} + g(0 - h) + \underbrace{\frac{2 f L_{\text{eq}} \langle v_2 \rangle^2}{d}}_{\widehat{E}_v} = 0, \qquad (13.6\text{-}4)$$

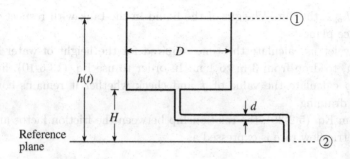

Figure 13.12   Draining of a tank with a pipe system.

where L_{eq} is the equivalent length of the drain pipe. Solving for $\langle v_2 \rangle$ gives

$$\langle v_2 \rangle^2 = \frac{2gh}{1 + \dfrac{4fL_{eq}}{d}}. \tag{13.6-5}$$

Considering the contents of the tank as a system, the conservation statement for mass becomes

$$- \text{Rate of mass out} = \text{Rate of mass accumulation}, \tag{13.6-6}$$

or

$$- \rho \langle v_2 \rangle \frac{\pi d^2}{4} = \frac{d}{dt} \left[\frac{\pi D^2}{4} (h - H)\rho \right]. \tag{13.6-7}$$

Simplification of Eq. (13.6-7) gives

$$- \langle v_2 \rangle \left(\frac{d}{D} \right)^2 = \frac{dh}{dt}. \tag{13.6-8}$$

Substitution of Eq. (13.6-5) into Eq. (13.6-8) and rearrangement give

$$dt = - \left(\frac{D}{d} \right)^2 \sqrt{\frac{1}{2g} \left(1 + \frac{4fL_{eq}}{d} \right)} \frac{dh}{\sqrt{h}}. \tag{13.6-9}$$

Analytical integration of Eq. (13.6-9) is possible only if the friction factor, f, is constant. In this case, the integration of Eq. (13.6-9) yields

$$t = \left(\frac{D}{d} \right)^2 \sqrt{\frac{2}{g} \left(1 + \frac{4fL_{eq}}{d} \right)} \left(\sqrt{h_o} - \sqrt{h} \right), \tag{13.6-10}$$

where h_o is the initial height of the liquid in the tank with respect to the reference plane.

Now let us calculate the time required for the height of water in the tank, h, to drop from 3 m to 1 m. In order to use Eq. (13.6-10), first we have to calculate the value of f and check whether it remains constant during draining.

From Eq. (5.2-39), the relationship between the friction factor and the volumetric flow rate is expressed as

$$f = \left(\frac{X}{Q}\right)^2,$$
(13.6-11)

where X is defined by

$$X = \sqrt{\frac{\pi^2 D^5 |\Delta\mathcal{P}|}{32\rho L}}.$$
(13.6-12)

At any instant, the pressure drop in the drain pipe system is $\rho g(h - H)$. Therefore, the term X takes the following form:

$$X = \sqrt{\frac{\pi^2 d^5 g(h - H)}{32 L_{eq}}}.$$
(13.6-13)

Note that the Reynolds number is defined by

$$\text{Re} = \frac{4\rho Q}{\pi \mu d}.$$
(13.6-14)

Substitution of Eqs. (13.6-11) and (13.6-14) into Eq. (13.6-3) leads to

$$Q = -4X\log\left(\frac{\varepsilon/d}{3.7065} + \frac{0.991\mu d}{\rho X}\right).$$
(13.6-15)

Taking the physical properties of water as

$$\rho = 1000 \text{ kg/m}^3 \quad \text{and} \quad \mu = 1 \times 10^{-3} \text{ Pa} \cdot \text{s},$$

the values of X, Q, and f as a function of the liquid height are calculated from Eqs. (13.6-13), (13.6-15), and (13.6-11), respectively, and tabulated as follows:

h (m)	$X \times 10^3$ (m^3/s)	Q (m^3/s)	$f \times 10^3$
3.0	5.498	0.088	3.890
2.5	4.917	0.078	3.926
2.0	4.259	0.068	3.977
1.5	3.477	0.055	4.059
1.0	2.459	0.038	4.226

Since the variation in f with the liquid height is negligible, we can assume that f remains almost constant at an arithmetic average value of 0.004.

Substitution of the numerical values into Eq. (13.6-10) gives

$$t = \left(\frac{5}{0.20}\right)^2 \sqrt{\frac{2}{9.8}} \left[1 + \frac{(4)(0.004)(80)}{0.20}\right] \left(\sqrt{3} - \sqrt{1}\right)$$

$$= 562 \text{ s} = 9.4 \text{ min.}$$

Comment: From Eq. (13.6-10),

$$\sqrt{h} = \sqrt{3} - \frac{t\,(d/D)^2}{\sqrt{\dfrac{2}{g}\left(1 + \dfrac{4fL_{eq}}{d}\right)}}.$$ (13.6-16)

The values of h, $\langle v_2 \rangle$, and dh/dt as a function of time are calculated from Eqs. (13.6-16), (13.6-5), and (13.6-8), respectively, and tabulated as follows:

t (s)	h (m)	$\langle v_2 \rangle$ (m/s)	dh/dt (m/s)
100	2.566	2.607	− 0.0042
200	2.166	2.395	− 0.0038
300	1.800	2.183	− 0.0035
400	1.467	1.971	− 0.0032
500	1.169	1.759	− 0.0028
562	1.000	1.628	− 0.0026

Note that dh/dt is negligible at all times in comparison with the liquid velocity through the drain pipe system.

Appendix A

Vector and Tensor Algebra

A.1 The Operations on Vectors

A vector is a quantity that associates a scalar with each coordinate direction. This definition implies that each vector has a magnitude and a direction. The magnitude of the vector $\mathbf{v}$ is designated by $|\mathbf{v}|$. Two vectors, $\mathbf{v}$ and $\mathbf{w}$, are said to be equal if $|\mathbf{v}| = |\mathbf{w}|$ and they point in the same direction. The vectors $\mathbf{v}$ and $-\mathbf{v}$ have the same magnitude $|\mathbf{v}|$, but point in opposite directions.

Addition of two vectors is carried out by placing them together so that the head of the first vector joins the tail of the second vector.

Multiplication of a vector $\mathbf{v}$ by a scalar α results in another vector of length $|\alpha|\,|\mathbf{v}|$; the direction of $\alpha\mathbf{v}$ is the same as that of $\mathbf{v}$ if $\alpha > 0$ and in the opposite direction if $\alpha < 0$.

The scalar (or dot) product of two vectors $\mathbf{v}$ and $\mathbf{w}$ is defined as the product of the magnitudes of the two vectors times the cosine of the angle between the vectors, i.e.,

$$\mathbf{v} \cdot \mathbf{w} = |\mathbf{v}|\,|\mathbf{w}|\cos\theta. \qquad (\text{A.1-1})$$

As the name implies, $\mathbf{v} \cdot \mathbf{w}$ is a scalar quantity. Consider the vectors $\mathbf{v}$ and $\mathbf{w}$ as shown in Figure A.1. The cosine of the angle between these two vectors can be expressed as

$$\cos\theta = \frac{\overline{OA}}{|\mathbf{w}|} = \frac{\overline{OB}}{|\mathbf{v}|}. \qquad (\text{A.1-2})$$

The use of Eq. (A.1-2) in Eq. (A.1-1) gives

$$\mathbf{v} \cdot \mathbf{w} = |\mathbf{v}|\,\overline{OA} = |\mathbf{w}|\,\overline{OB}. \qquad (\text{A.1-3})$$

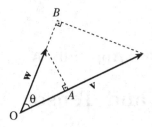

Figure A.1   The scalar product of two vectors.

Note that

$$\overline{OA} = \text{Projection of } \mathbf{w} \text{ in the direction of } \mathbf{v},$$
$$\overline{OB} = \text{Projection of } \mathbf{v} \text{ in the direction of } \mathbf{w}.$$

Thus, Eq. (A.1-3) implies that the scalar product is the magnitude of $\mathbf{v}$ multiplied by the projection of $\mathbf{w}$ on $\mathbf{v}$ or the magnitude of $\mathbf{w}$ multiplied by the projection of $\mathbf{v}$ on $\mathbf{w}$. The scalar product of a vector with itself is expressed as

$$\mathbf{v} \cdot \mathbf{v} = |\mathbf{v}|^2 \quad \Rightarrow \quad |\mathbf{v}| = \sqrt{\mathbf{v} \cdot \mathbf{v}}. \tag{A.1-4}$$

The vector (or cross) product of two vectors $\mathbf{v}$ and $\mathbf{w}$ is a vector defined by

$$\mathbf{v} \times \mathbf{w} = |\mathbf{v}|\,|\mathbf{w}| \sin \theta \, \mathbf{n}, \tag{A.1-5}$$

where $\mathbf{n}$ is a vector of unit length. The direction of $\mathbf{n}$ is determined by the right-hand rule, i.e., when the index finger points in the direction of $\mathbf{v}$ and the middle finger points in the direction of $\mathbf{w}$, the thumb points in the direction of $\mathbf{n}$. The magnitude of the vector $\mathbf{v} \times \mathbf{w}$ is the area of the parallelogram formed by the vectors $\mathbf{v}$ and $\mathbf{w}$.

A.2   Basis and Basis Vectors

Any vector $\mathbf{v}$ can be written as a linear combination of $\mathbf{e}_1$, $\mathbf{e}_2$, $\mathbf{e}_3$ such that

$$\mathbf{v} = v_1\mathbf{e}_1 + v_2\mathbf{e}_2 + v_3\mathbf{e}_3 = \sum_{i=1}^{3} v_i\mathbf{e}_i, \tag{A.2-1}$$

where v_1, v_2, v_3 are called the *components* of the vector $\mathbf{v}$, and $\mathbf{e}_1$, $\mathbf{e}_2$, $\mathbf{e}_3$ are called the *basis vectors*.

A basis $(\mathbf{e}_1, \mathbf{e}_2, \mathbf{e}_3)$ is said to be *orthonormal* if each element in the set is a unit vector, i.e.,

$$\mathbf{e}_1 \cdot \mathbf{e}_1 = \mathbf{e}_2 \cdot \mathbf{e}_2 = \mathbf{e}_3 \cdot \mathbf{e}_3 = 1 \qquad (A.2\text{-}2)$$

and the elements are orthogonal to one another at every point, i.e.,

$$\mathbf{e}_i \cdot \mathbf{e}_j = 0 \quad \text{for } i \neq j. \qquad (A.2\text{-}3)$$

The basis vectors $(\mathbf{e}_1, \mathbf{e}_2, \mathbf{e}_3)$ in the Cartesian coordinate system are orthonormal, i.e., they are of unit length and mutually orthogonal.[1] Moreover, they point in the same direction at two different locations. This implies

$$\mathbf{e}_i(x, y, z) = \mathbf{e}_i(x^*, y^*, z^*) \quad \text{for } i = 1, 2, 3. \qquad (A.2\text{-}4)$$

In other words, Cartesian basis vectors are independent of position.

In Cartesian coordinates, each vector can be written as a linear combination of the basis vectors as

$$\mathbf{v} = v_x \mathbf{e}_x + v_y \mathbf{e}_y + v_z \mathbf{e}_z. \qquad (A.2\text{-}5)$$

As shown in Figure A.2, v_x, v_y, and v_z are the components of vector $\mathbf{v}$. They are simply the projections of the vector $\mathbf{v}$ along the coordinate axes x, y, and z, i.e.,

$$v_x = \overline{OA} = \mathbf{v} \cdot \mathbf{e}_x \qquad v_y = \overline{OB} = \mathbf{v} \cdot \mathbf{e}_y \qquad v_z = \overline{OC} = \mathbf{v} \cdot \mathbf{e}_z. \qquad (A.2\text{-}6)$$

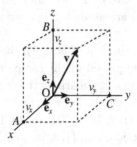

Figure A.2 The projections of a vector on the Cartesian coordinate axes x, y, and z.

---

[1] In physics textbooks, $\mathbf{e}_1$, $\mathbf{e}_2$, and $\mathbf{e}_3$ are replaced by $\boldsymbol{i}$, $\boldsymbol{j}$, and $\boldsymbol{k}$, respectively.

A.3   Summation Convention

When a subscript or superscript (index) is repeated in a given term of an expression, then summation over that index is implied. Using the summation convention, Eq. (A.2-1) can be expressed as

$$\mathbf{v} = v_i \mathbf{e}_i \qquad (A.3-1)$$

with the understanding that i goes from 1 to 3 since it is repeated. A repeated index is called a *dummy index* because the expression is independent of the letter used for the repeated index. An index that is not repeated is called a *free index*.

The scalar product of two basis vectors of unit length is given by

$$\mathbf{e}_i \cdot \mathbf{e}_j = \delta_{ij}, \qquad (A.3-2)$$

where δ_{ij}, the *Kronecker delta*, is defined by

$$\delta_{ij} = \begin{cases} 1 & \text{if } i = j \\ 0 & \text{if } i \neq j \end{cases}. \qquad (A.3-3)$$

On the other hand, the vector product of two basis vectors of unit length is given by

$$\mathbf{e}_i \times \mathbf{e}_j = \epsilon_{ijk}\, \mathbf{e}_k, \qquad (A.3-4)$$

where ϵ_{ijk} is called the *alternating unit tensor* (or *permutation symbol*). It is defined as

$$\epsilon_{ijk} = \begin{cases} 0 & \text{when any two of the indices are equal} \\ +1 & \text{when } (i,j,k) \text{ are } (1,2,3) \text{ or an even permutation of } (1,2,3) \\ -1 & \text{when } (i,j,k) \text{ are an odd permutation of } (1,2,3) \end{cases}. \qquad (A.3-5)$$

If (i, j, k) are an even permutation of $(1, 2, 3)$, then (i, j, k) require an even number of inversions to be placed in the same order as $(1, 2, 3)$. The sign convention can be remembered by marking the numbers $(1, 2, 3)$ on a circle. Then any even permutation will be in cyclic order, i.e., it will go around the circle in the same sense as $(1, 2, 3)$, $(2, 3, 1)$, or $(3, 1, 2)$. An odd permutation will go in the reverse direction, i.e., $(1, 3, 2)$, $(3, 2, 1)$, or $(2, 1, 3)$.

The following relations involving the Kronecker delta and alternating unit tensor are useful in proving some vector identities:

$$\epsilon_{ijk}\, \epsilon_{mjk} = 2\, \delta_{im}, \qquad (A.3-6)$$

$$\epsilon_{ijk}\, \epsilon_{mnk} = \delta_{im}\, \delta_{jn} - \delta_{in}\, \delta_{jm}. \qquad (A.3-7)$$

Using summation convention, various operations on vectors are carried out as follows:

- Vector addition

$$\mathbf{v} + \mathbf{w} = v_i \mathbf{e}_i + w_i \mathbf{e}_i = (v_i + w_i)\,\mathbf{e}_i. \tag{A.3-8}$$

- Scalar multiplication

$$\alpha\,\mathbf{v} = (\alpha v_i)\,\mathbf{e}_i. \tag{A.3-9}$$

- Scalar product

$$\mathbf{v} \cdot \mathbf{w} = (v_i \mathbf{e}_i) \cdot (w_j \mathbf{e}_j) = v_i w_j \underbrace{(\mathbf{e}_i \cdot \mathbf{e}_j)}_{\delta_{ij}} = v_i w_i. \tag{A.3-10}$$

- Vector product

$$\mathbf{v} \times \mathbf{w} = (v_i \mathbf{e}_i) \times (w_j \mathbf{e}_j) = v_i w_j (\mathbf{e}_i \times \mathbf{e}_j) = \epsilon_{ijk} v_i w_j \mathbf{e}_k. \tag{A.3-11}$$

Summation over the indices i, j, and k leads to 27 terms. Equation (A.3-11) can also be expressed in the form of a determinant as

$$\mathbf{v} \times \mathbf{w} = \begin{vmatrix} \mathbf{e}_1 & \mathbf{e}_2 & \mathbf{e}_3 \\ v_1 & v_2 & v_3 \\ w_1 & w_2 & w_3 \end{vmatrix}. \tag{A.3-12}$$

Example 1. Show that

$$\mathbf{v} \cdot (\mathbf{v} \times \mathbf{w}) = 0.$$

Solution. The left-hand side is expressed as

$$\begin{aligned} \mathbf{v} \cdot (\mathbf{v} \times \mathbf{w}) &= v_i \mathbf{e}_i \cdot (v_j \mathbf{e}_j \times w_m \mathbf{e}_m) \\ &= v_i \mathbf{e}_i \cdot v_j w_m \epsilon_{jmk} \mathbf{e}_k \\ &= \epsilon_{jmk} v_i v_j w_m \delta_{ik} \\ &= \epsilon_{jmi} v_i v_j w_m. \end{aligned}$$

If we expand the resultant term on j,

$$\mathbf{v} \cdot (\mathbf{v} \times \mathbf{w}) = \epsilon_{1mi} v_i v_1 w_m + \epsilon_{2mi} v_i v_2 w_m + \epsilon_{3mi} v_i v_3 w_m.$$

Expanding on m,

$$\begin{aligned} \mathbf{v} \cdot (\mathbf{v} \times \mathbf{w}) = &\,\epsilon_{12i} v_i v_1 w_2 + \epsilon_{13i} v_i v_1 w_3 + \epsilon_{21i} v_i v_2 w_1 + \epsilon_{23i} v_i v_2 w_3 \\ &+ \epsilon_{31i} v_i v_3 w_1 + \epsilon_{32i} v_i v_3 w_2. \end{aligned}$$

Expanding on i,

$$\mathbf{v} \cdot (\mathbf{v} \times \mathbf{w}) = \epsilon_{123} v_3 v_1 w_2 + \epsilon_{132} v_2 v_1 w_3 + \epsilon_{213} v_3 v_2 w_1 + \epsilon_{231} v_1 v_2 w_3$$
$$+ \epsilon_{312} v_2 v_3 w_1 + \epsilon_{321} v_1 v_3 w_2,$$

or

$$\mathbf{v} \cdot (\mathbf{v} \times \mathbf{w}) = v_3 v_1 w_2 - v_2 v_1 w_3 - v_3 v_2 w_1 + v_1 v_2 w_3 + v_2 v_3 w_1 - v_1 v_3 w_2$$
$$= 0.$$

Example 2.   Show that

$$(\mathbf{u} \times \mathbf{v}) \cdot (\mathbf{w} \times \mathbf{z}) = (\mathbf{u} \cdot \mathbf{w})(\mathbf{v} \cdot \mathbf{z}) - (\mathbf{u} \cdot \mathbf{z})(\mathbf{v} \cdot \mathbf{w}).$$

Solution.   Using summation convention, the left-hand side of the given equation is expressed as

$$
\begin{aligned}
(\mathbf{u} \times \mathbf{v}) \cdot (\mathbf{w} \times \mathbf{z}) &= (u_i \mathbf{e}_i \times v_j \mathbf{e}_j) \cdot (w_m \mathbf{e}_m \times z_n \mathbf{e}_n) \\
&= [u_i v_j (\mathbf{e}_i \times \mathbf{e}_j)] \cdot [w_m z_n (\mathbf{e}_m \times \mathbf{e}_n)] \\
&= (\epsilon_{ijk} u_i v_j \mathbf{e}_k) \cdot (\epsilon_{mnr} w_m z_n \mathbf{e}_r) \\
&= \epsilon_{ijk} \epsilon_{mnr} u_i v_j w_m z_n (\mathbf{e}_k \cdot \mathbf{e}_r) \\
&= \epsilon_{ijk} \epsilon_{mnr} u_i v_j w_m z_n \delta_{kr}.
\end{aligned}
$$

$$
\begin{aligned}
(\mathbf{u} \times \mathbf{v}) \cdot (\mathbf{w} \times \mathbf{z}) &= \epsilon_{ijr} \epsilon_{mnr} u_i v_j w_m z_n \\
&= \underbrace{\delta_{im} \delta_{jn} u_i v_j w_m z_n}_{i=m \ \& \ j=n} - \underbrace{\delta_{in} \delta_{jm} u_i v_j w_m z_n}_{i=n \ \& \ j=m} \\
&= (u_i w_i)(v_j z_j) - (u_i z_i)(v_j w_j) \\
&= (\mathbf{u} \cdot \mathbf{w})(\mathbf{v} \cdot \mathbf{z}) - (\mathbf{u} \cdot \mathbf{z})(\mathbf{v} \cdot \mathbf{w}).
\end{aligned}
$$

A.4   Second-Order Tensor

A second-order tensor is a quantity that associates a vector with each coordinate direction or a quantity that associates a scalar with each ordered pair of coordinate directions. Using the summation convention, a second-order tensor $\mathbf{T}$ is defined by

$$\mathbf{T} = T_{ij} \mathbf{e}_i \mathbf{e}_j, \tag{A.4-1}$$

where the T_{ij}s are called the *components* of the tensor $\mathbf{T}$ and can be displayed as a matrix in the form

$$T_{ij} = \begin{pmatrix} T_{11} & T_{12} & T_{13} \\ T_{21} & T_{22} & T_{23} \\ T_{31} & T_{32} & T_{33} \end{pmatrix}. \tag{A.4-2}$$

When two vector quantities are written side by side without any multiplication sign between them, the result is a second-order tensor called a *dyadic product*. The term $\mathbf{e}_i\mathbf{e}_j$ in Eq. (A.4-1) is a dyadic product of basis vectors of unit length.

A second-order tensor $\mathbf{T}$ transforms a vector into another vector such that

$$\mathbf{T} \cdot \mathbf{v} = (T_{ij}\mathbf{e}_i\mathbf{e}_j) \cdot (v_k\mathbf{e}_k) = T_{ij}v_k\underbrace{(\mathbf{e}_j \cdot \mathbf{e}_k)}_{\delta_{jk}}\mathbf{e}_i = (T_{ik}v_k)\mathbf{e}_i. \tag{A.4-3}$$

On the other hand,

$$\mathbf{v} \cdot \mathbf{T} = (v_k\mathbf{e}_k) \cdot (T_{ij}\mathbf{e}_i\mathbf{e}_j) = T_{ij}v_k\underbrace{(\mathbf{e}_k \cdot \mathbf{e}_i)}_{\delta_{ki}}\mathbf{e}_j = (T_{ij}v_i)\mathbf{e}_j. \tag{A.4-4}$$

Thus, $\mathbf{T} \cdot \mathbf{v} \neq \mathbf{v} \cdot \mathbf{T}$.

The following are several special kinds of second-order tensors encountered in transport phenomena:

- A *unit tensor* or an *identity tensor* $\mathbf{I}$, defined by

$$\mathbf{I} = \delta_{ij}\mathbf{e}_i\mathbf{e}_j, \tag{A.4-5}$$

  transforms every vector into itself, i.e.,

$$\mathbf{I} \cdot \mathbf{v} = \mathbf{v} \cdot \mathbf{I} = \mathbf{v}. \tag{A.4-6}$$

- The *transpose* of $\mathbf{T}$, $\mathbf{T}^{\mathrm{T}}$, is defined as follows:

$$\mathbf{T} \cdot \mathbf{v} = \mathbf{v} \cdot \mathbf{T}^{\mathrm{T}}. \tag{A.4-7}$$

If a second-order tensor $\mathbf{T}$ and its components are defined as

$$\mathbf{T} = T_{ij}\mathbf{e}_i\mathbf{e}_j \text{ with } T_{ij} = \begin{pmatrix} T_{11} & T_{12} & T_{13} \\ T_{21} & T_{22} & T_{23} \\ T_{31} & T_{32} & T_{33} \end{pmatrix}, \tag{A.4-8}$$

then the transpose of $\mathbf{T}$, $\mathbf{T}^{\mathrm{T}}$, is given by

$$\mathbf{T}^{\mathrm{T}} = T_{ji}\mathbf{e}_i\mathbf{e}_j \text{ with } T_{ji} = \begin{pmatrix} T_{11} & T_{21} & T_{31} \\ T_{12} & T_{22} & T_{32} \\ T_{13} & T_{23} & T_{33} \end{pmatrix}. \tag{A.4-9}$$

If $\mathbf{T}$ and $\mathbf{U}$ are any two second-order tensors, then

$$(\mathbf{T} \cdot \mathbf{U})^{\mathrm{T}} = \mathbf{U}^{\mathrm{T}} \cdot \mathbf{T}^{\mathrm{T}}. \tag{A.4-10}$$

- A second-order tensor $\mathbf{T}$ is said to be *symmetric* if

$$\mathbf{T} = \mathbf{T}^{\mathrm{T}} \quad \text{or} \quad T_{ij} = T_{ji}. \tag{A.4-11}$$

Therefore, the number of independent components of a symmetric tensor is 6.

- A second-order tensor is said to be *skew-symmetric* (or antisymmetric) if

$$\mathbf{T} = -\mathbf{T}^{\mathrm{T}} \quad \text{or} \quad T_{ij} = -T_{ji}. \tag{A.4-12}$$

Since the diagonal elements (T_{ii}) are all zero according to Eq. (A.4-12), the number of independent components of a skew-symmetric tensor is 3.

A.5   Vector and Tensor Differential Operations

The vector differential operator ∇, known as *del* or *nabla*, is defined in the Cartesian coordinate system as

$$\nabla = \frac{\partial}{\partial x_i}\mathbf{e}_i = \frac{\partial}{\partial x}\mathbf{e}_x + \frac{\partial}{\partial y}\mathbf{e}_y + \frac{\partial}{\partial z}\mathbf{e}_z. \tag{A.5-1}$$

Keep in mind that Cartesian basis vectors are independent of position.

The *gradient of a scalar field* α is a vector field denoted by $\nabla\alpha$. In the Cartesian coordinate system, it is expressed as

$$\nabla\alpha = \frac{\partial\alpha}{\partial x_i}\mathbf{e}_i = \frac{\partial\alpha}{\partial x}\mathbf{e}_x + \frac{\partial\alpha}{\partial y}\mathbf{e}_y + \frac{\partial\alpha}{\partial z}\mathbf{e}_z. \tag{A.5-2}$$

The components of $\nabla\alpha$ represent the rate of change of the scalar field with respect to each coordinate direction.

The *divergence of a vector field* $\mathbf{v}$ is a scalar field denoted by $\nabla \cdot \mathbf{v}$. In the Cartesian coordinate system, it is expressed as

$$\nabla \cdot \mathbf{v} = \frac{\partial}{\partial x_i}\mathbf{e}_i \cdot (v_j\mathbf{e}_j) = \frac{\partial v_j}{\partial x_i}\underbrace{(\mathbf{e}_i \cdot \mathbf{e}_j)}_{\delta_{ij}} = \frac{\partial v_i}{\partial x_i}$$

$$= \frac{\partial v_x}{\partial x} + \frac{\partial v_y}{\partial y} + \frac{\partial v_z}{\partial z}. \tag{A.5-3}$$

The *curl of a vector field* **v** is also a vector field denoted by $\nabla \times \mathbf{v}$. In the Cartesian coordinate system, the vector product between the ∇ operator and the vector **v** is expressed as

$$\nabla \times \mathbf{v} = \frac{\partial}{\partial x_i}\, \mathbf{e}_i \times (v_j \mathbf{e}_j) = \frac{\partial v_j}{\partial x_i}(\mathbf{e}_i \times \mathbf{e}_j) = \epsilon_{ijk}\, \frac{\partial v_j}{\partial x_i}\, \mathbf{e}_k. \tag{A.5-4}$$

Summation over the indices i, j, and k leads to 27 terms. Equation (A.5-4) can also be expressed in the form of a determinant as

$$\nabla \times \mathbf{v} = \begin{vmatrix} \mathbf{e}_x & \mathbf{e}_y & \mathbf{e}_z \\ \dfrac{\partial}{\partial x} & \dfrac{\partial}{\partial y} & \dfrac{\partial}{\partial z} \\ v_x & v_y & v_z \end{vmatrix}. \tag{A.5-5}$$

The *gradient of a vector field* **v** is a second-order tensor field denoted by $\nabla \mathbf{v}$. In the Cartesian coordinate system, it is expressed as

$$\nabla \mathbf{v} = \frac{\partial v_j}{\partial x_i}\, \mathbf{e}_i \mathbf{e}_j. \tag{A.5-6}$$

The transpose of $\nabla \mathbf{v}$ is

$$(\nabla \mathbf{v})^{\mathrm{T}} = \frac{\partial v_i}{\partial x_j}\, \mathbf{e}_i \mathbf{e}_j. \tag{A.5-7}$$

The *Laplacian of a scalar field* α is the divergence of the gradient of a scalar field α. In the Cartesian coordinate system, it is expressed as

$$\nabla \cdot \nabla \alpha = \nabla^2 \alpha = \frac{\partial}{\partial x_i}\, \mathbf{e}_i \cdot \frac{\partial \alpha}{\partial x_j}\, \mathbf{e}_j = \frac{\partial^2 \alpha}{\partial x_i\, \partial x_j}\, \underbrace{(\mathbf{e}_i \cdot \mathbf{e}_j)}_{\delta_{ij}} = \frac{\partial^2 \alpha}{\partial x_i\, \partial x_i}$$

$$= \frac{\partial^2 \alpha}{\partial x^2} + \frac{\partial^2 \alpha}{\partial y^2} + \frac{\partial^2 \alpha}{\partial z^2}. \tag{A.5-8}$$

In Eq. (A.5-8), the term $\nabla \cdot \nabla = \nabla^2$ is called the *Laplacian* operator.

Some of the useful identities involving vector and tensor differential operations are as follows:

$$\nabla \cdot \alpha\, \mathbf{v} = \alpha\, (\nabla \cdot \mathbf{v}) + \mathbf{v} \cdot \nabla \alpha, \tag{A.5-9}$$

$$\nabla \cdot \alpha\, \mathbf{I} = \nabla \alpha, \tag{A.5-10}$$

$$\nabla \cdot \mathbf{vw} = \mathbf{v} \cdot \nabla \mathbf{w} + \mathbf{w}\, (\nabla \cdot \mathbf{v}), \tag{A.5-11}$$

$$\nabla \times \alpha\, \mathbf{v} = \nabla \alpha \times \mathbf{v} + \alpha\, (\nabla \times \mathbf{v}), \tag{A.5-12}$$

$$\nabla \cdot \alpha\, \mathbf{T} = \nabla \alpha \cdot \mathbf{T} + \alpha\, (\nabla \cdot \mathbf{T}). \tag{A.5-13}$$

Example 3. If $\alpha = x^2 yz$ and $\beta = xy - 3z^2$, find (a) $\nabla \cdot (\nabla\alpha \times \nabla\beta)$, (b) $\nabla \times (\nabla\alpha \times \nabla\beta)$.

Solution. The gradients of α and β are

$$\nabla\alpha = \frac{\partial(x^2yz)}{\partial x}\mathbf{e}_x + \frac{\partial(x^2yz)}{\partial y}\mathbf{e}_y + \frac{\partial(x^2yz)}{\partial z}\mathbf{e}_z$$

$$= 2xyz\,\mathbf{e}_x + x^2z\,\mathbf{e}_y + x^2y\,\mathbf{e}_z,$$

$$\nabla\beta = \frac{\partial(xy - 3z^2)}{\partial x}\mathbf{e}_x + \frac{\partial(xy - 3z^2)}{\partial y}\mathbf{e}_y + \frac{\partial(xy - 3z^2)}{\partial z}\mathbf{e}_z$$

$$= y\,\mathbf{e}_x + x\,\mathbf{e}_y - 6z\,\mathbf{e}_z.$$

Thus, the use of Eq. (A.3-12) gives

$$\nabla\alpha \times \nabla\beta = \begin{vmatrix} \mathbf{e}_x & \mathbf{e}_y & \mathbf{e}_z \\ 2xyz & x^2z & x^2y \\ y & x & -6z \end{vmatrix}$$

$$= -(6x^2z^2 + x^3y)\mathbf{e}_x + (12xyz^2 + x^2y^2)\mathbf{e}_y + x^2yz\,\mathbf{e}_z.$$

(a) The use of Eq. (A.5-3) gives

$$\nabla \cdot (\nabla\alpha \times \nabla\beta) = -\frac{\partial(6x^2z^2 + x^3y)}{\partial x} + \frac{\partial(12xyz^2 + x^2y^2)}{\partial y} + \frac{\partial(x^2yz)}{\partial z}$$

$$= -12xz^2 - 3x^2y + 12xz^2 + 2x^2y + x^2y = 0.$$

(b) The use of Eq. (A.5-5) gives

$$\nabla \times (\nabla\alpha \times \nabla\beta) = \begin{vmatrix} \mathbf{e}_x & \mathbf{e}_y & \mathbf{e}_z \\ \dfrac{\partial}{\partial x} & \dfrac{\partial}{\partial y} & \dfrac{\partial}{\partial z} \\ -(6x^2z^2 + x^3y) & 12xyz^2 + x^2y^2v_y & x^2yz \end{vmatrix}$$

$$= (x^2z - 24xyz)\mathbf{e}_x - (2xyz + 12x^2z)\mathbf{e}_y$$

$$+ (12yz^2 + 2xy^2 + x^3)\mathbf{e}_z.$$

Example 4. Show that

$$\nabla \cdot \alpha\mathbf{T} = \nabla\alpha \cdot \mathbf{T} + \alpha\nabla \cdot \mathbf{T}.$$

Solution. The term on the left-hand side is expressed as

$$\nabla \cdot \alpha \mathbf{T} = \frac{\partial}{\partial x_i} \mathbf{e}_i \cdot \alpha T_{mn} \mathbf{e}_m \mathbf{e}_n = \frac{\partial (\alpha T_{mn})}{\partial x_i} \underbrace{(\mathbf{e}_i \cdot \mathbf{e}_m)}_{\delta_{im}} \mathbf{e}_n$$

$$= \alpha \frac{\partial T_{mn}}{\partial x_m} \mathbf{e}_n + T_{mn} \frac{\partial \alpha}{\partial x_m} \mathbf{e}_n.$$

On the other hand, the terms on the right-hand side are expressed as

$$\nabla \alpha \cdot \mathbf{T} = \frac{\partial \alpha}{\partial x_i} \mathbf{e}_i \cdot T_{mn} \mathbf{e}_m \mathbf{e}_n = \frac{\partial \alpha}{\partial x_i} T_{mn} \underbrace{(\mathbf{e}_i \cdot \mathbf{e}_m)}_{\delta_{im}} \mathbf{e}_n = \frac{\partial \alpha}{\partial x_m} T_{mn} \mathbf{e}_n,$$

$$\alpha \nabla \cdot \mathbf{T} = \alpha \frac{\partial}{\partial x_i} \mathbf{e}_i \cdot T_{mn} \mathbf{e}_m \mathbf{e}_n = \alpha \frac{\partial T_{mn}}{\partial x_i} \underbrace{(\mathbf{e}_i \cdot \mathbf{e}_m)}_{\delta_{im}} \mathbf{e}_n = \alpha \frac{\partial T_{mn}}{\partial x_m} \mathbf{e}_n.$$

Therefore, the summation of the terms on the right-hand side of the given expression is equivalent to the term on the left-hand side.

A.6 Vector and Tensor Algebra in Curvilinear Coordinates

Besides Cartesian coordinates, the two most commonly used coordinate systems are *cylindrical* and *spherical coordinates*. In these two curvilinear coordinate systems, however, the basis vectors are dependent on position.

The cylindrical coordinate system is shown in Figure A.3. For cylindrical coordinates, the variables (r, θ, z) are related to the Cartesian coordinates (x, y, z) as follows:

$$\begin{aligned} x &= r\cos\theta & r &= \sqrt{x^2 + y^2}, \\ y &= r\sin\theta & \theta &= \arctan(y/x), \\ z &= z & z &= z. \end{aligned} \tag{A.6-1}$$

The ranges of the variables (r, θ, z) are

$$0 \leq r \leq \infty \qquad 0 \leq \theta \leq 2\pi \qquad -\infty \leq z \leq \infty. \tag{A.6-2}$$

The differential volume and areas in the cylindrical coordinate system are given as follows:

$$dV = r\, dr\, d\theta\, dz. \tag{A.6-3}$$

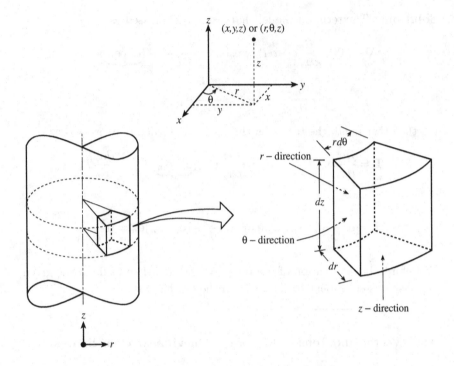

Figure A.3   The cylindrical coordinate system.

$$dA = \begin{cases} r\,d\theta\,dz & \text{Flux is in the } r\text{-direction,} \\ dr\,dz & \text{Flux is in the } \theta\text{-direction,} \\ r\,dr\,d\theta & \text{Flux is in the } z\text{-direction.} \end{cases} \qquad (A.6\text{-}4)$$

The relationships between the basis vectors of the Cartesian and cylindrical coordinate systems are expressed in the form

$$\mathbf{e}_x = \cos\theta\,\mathbf{e}_r - \sin\theta\,\mathbf{e}_\theta, \qquad (A.6\text{-}5)$$

$$\mathbf{e}_y = \sin\theta\,\mathbf{e}_r + \cos\theta\,\mathbf{e}_\theta, \qquad (A.6\text{-}6)$$

$$\mathbf{e}_z = \mathbf{e}_z. \qquad (A.6\text{-}7)$$

and

$$\mathbf{e}_r = \cos\theta\,\mathbf{e}_x + \sin\theta\,\mathbf{e}_y, \qquad (A.6\text{-}8)$$

$$\mathbf{e}_\theta = -\sin\theta\,\mathbf{e}_x + \cos\theta\,\mathbf{e}_y, \qquad (A.6\text{-}9)$$

$$\mathbf{e}_z = \mathbf{e}_z. \qquad (A.6\text{-}10)$$

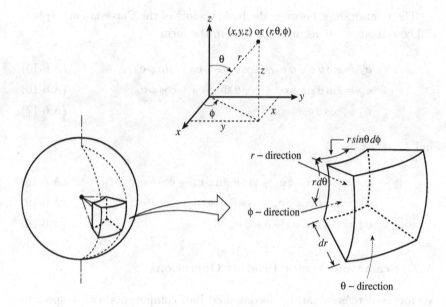

Figure A.4 The spherical coordinate system.

The spherical coordinate system is shown in Figure A.4. For spherical coordinates, the variables (r, θ, ϕ) are related to the Cartesian coordinates (x, y, z) as follows:

$$
\begin{aligned}
x &= r \sin \theta \cos \phi & r &= \sqrt{x^2 + y^2 + z^2}, \\
y &= r \sin \theta \sin \phi & \theta &= \arctan\left(\sqrt{x^2 + y^2}/z\right), \\
z &= r \cos \theta & \phi &= \arctan(y/x).
\end{aligned}
\tag{A.6-11}
$$

The ranges of the variables (r, θ, ϕ) are

$$
0 \leq r \leq \infty \qquad 0 \leq \theta \leq \pi \qquad 0 \leq \phi \leq 2\pi.
\tag{A.6-12}
$$

The differential volume and areas in the spherical coordinate system are as follows:

$$
dV = r^2 \sin \theta \, dr \, d\theta \, d\phi.
\tag{A.6-13}
$$

$$
dA = \begin{cases}
r^2 \sin \theta \, d\theta \, d\phi & \text{Flux is in the } r\text{-direction,} \\
r \sin \theta \, dr \, d\phi & \text{Flux is in the } \theta\text{-direction, .} \\
r \, dr \, d\theta & \text{Flux is in the } \phi\text{-direction}
\end{cases}
\tag{A.6-14}
$$

The relationships between the basis vectors of the Cartesian and spherical coordinate systems are expressed in the form

$$\mathbf{e}_x = \sin\theta\cos\phi\,\mathbf{e}_r + \cos\theta\cos\phi\,\mathbf{e}_\theta - \sin\phi\,\mathbf{e}_\phi, \qquad (A.6\text{-}15)$$

$$\mathbf{e}_y = \sin\theta\sin\phi\,\mathbf{e}_r + \cos\theta\sin\phi\,\mathbf{e}_\theta + \cos\phi\,\mathbf{e}_\phi, \qquad (A.6\text{-}16)$$

$$\mathbf{e}_z = \cos\theta\,\mathbf{e}_r - \sin\theta\,\mathbf{e}_\theta, \qquad (A.6\text{-}17)$$

and

$$\mathbf{e}_r = \sin\theta\cos\phi\,\mathbf{e}_x + \sin\theta\sin\phi\,\mathbf{e}_y + \cos\theta\,\mathbf{e}_z, \qquad (A.6\text{-}18)$$

$$\mathbf{e}_\theta = \cos\theta\cos\phi\,\mathbf{e}_x + \cos\theta\sin\phi\,\mathbf{e}_y - \sin\theta\,\mathbf{e}_z, \qquad (A.6\text{-}19)$$

$$\mathbf{e}_\phi = -\sin\phi\,\mathbf{e}_x + \cos\phi\,\mathbf{e}_y. \qquad (A.6\text{-}20)$$

A.7   Scalar and Vector Product Operations

Vectors and tensors can be decomposed into components with respect to curvilinear coordinates just as with respect to Cartesian coordinates, and the various scalar and vector product operations are performed in a similar way.

A.7.1   *Scalar product of two vectors*

The scalar product of two vectors is given by Eq. (A.3-10), i.e.,

$$\mathbf{v}\cdot\mathbf{w} = v_i w_i = v_1 w_1 + v_2 w_2 + v_3 w_3. \qquad (A.7\text{-}1)$$

- Cartesian coordinate system

$$\mathbf{v}\cdot\mathbf{w} = v_x w_x + v_y w_y + v_z w_z. \qquad (A.7\text{-}2)$$

- Cylindrical coordinate system

$$\mathbf{v}\cdot\mathbf{w} = v_r w_r + v_\theta w_\theta + v_z w_z. \qquad (A.7\text{-}3)$$

- Spherical coordinate system

$$\mathbf{v}\cdot\mathbf{w} = v_r w_r + v_\theta w_\theta + v_\phi w_\phi. \qquad (A.7\text{-}4)$$

A.7.2   *Vector product of two vectors*

The vector product of two vectors is given by Eq. (A.3-11), i.e.,

$$\mathbf{v} \times \mathbf{w} = (v_2 w_3 - v_3 w_2)\mathbf{e}_1 - (v_1 w_3 - v_3 w_1)\mathbf{e}_2 + (v_1 w_2 - v_2 w_1)\mathbf{e}_3. \quad \text{(A.7-5)}$$

- Cartesian coordinate system

$$\mathbf{v} \times \mathbf{w} = (v_y w_z - v_z w_y)\mathbf{e}_x - (v_x w_z - v_z w_x)\mathbf{e}_y + (v_x w_y - v_y w_x)\mathbf{e}_z. \quad \text{(A.7-6)}$$

- Cylindrical coordinate system

$$\mathbf{v} \times \mathbf{w} = (v_\theta w_z - v_z w_\theta)\mathbf{e}_r - (v_r w_z - v_z w_r)\mathbf{e}_\theta + (v_r w_\theta - v_\theta w_r)\mathbf{e}_z. \quad \text{(A.7-7)}$$

- Spherical coordinate system

$$\mathbf{v} \times \mathbf{w} = (v_\theta w_\phi - v_\phi w_\theta)\mathbf{e}_r - (v_r w_\phi - v_\phi w_r)\mathbf{e}_\theta + (v_r w_\theta - v_\theta w_r)\mathbf{e}_\phi. \quad \text{(A.7-8)}$$

A.7.3   *Double dot product of two second-order tensors*

The double dot product of two second-order tensors results in a scalar quantity. Two second-order tensors may be multiplied according to the double dot operation as follows:

$$\begin{aligned} \boldsymbol{\sigma} : \mathbf{T} &= \sigma_{ij}\mathbf{e}_i\mathbf{e}_j : T_{mn}\mathbf{e}_m\mathbf{e}_n = \sigma_{ij}T_{mn}(\mathbf{e}_i \cdot \mathbf{e}_n)(\mathbf{e}_j \cdot \mathbf{e}_m) \\ &= \sigma_{ij}T_{mn}\delta_{jm}\delta_{in} = \sigma_{ij}T_{ji}. \end{aligned} \quad \text{(A.7-9)}$$

Similarly, one can show that

$$\mathbf{T} : \mathbf{vw} = T_{ij}v_j w_i, \quad \text{(A.7-10)}$$

$$\mathbf{uv} : \mathbf{wz} = u_i v_j w_j z_i. \quad \text{(A.7-11)}$$

Also note that

$$\mathbf{w} \cdot \mathbf{v} \cdot \mathbf{T} = \mathbf{v} \cdot \mathbf{T} \cdot \mathbf{w} = \mathbf{wv} : \mathbf{T}. \quad \text{(A.7-12)}$$

A.8   Differential Operations

A.8.1   *Gradient of a scalar field*

In Cartesian coordinates, the gradient of a scalar field α is defined by Eq. (A.5-2), i.e.,

$$\nabla\alpha = \frac{\partial\alpha}{\partial x}\mathbf{e}_x + \frac{\partial\alpha}{\partial y}\mathbf{e}_y + \frac{\partial\alpha}{\partial z}\mathbf{e}_z. \quad \text{(A.8-1)}$$

In cylindrical and spherical coordinates, basis vectors are dependent on coordinate direction. Thus, the gradient of a scalar field α in these coordinate systems is as follows:

- Cylindrical coordinate system

$$\nabla \alpha = \frac{\partial \alpha}{\partial r} \mathbf{e}_r + \frac{1}{r} \frac{\partial \alpha}{\partial \theta} \mathbf{e}_\theta + \frac{\partial \alpha}{\partial z} \mathbf{e}_z. \qquad \text{(A.8-2)}$$

- Spherical coordinate system

$$\nabla \alpha = \frac{\partial \alpha}{\partial r} \mathbf{e}_r + \frac{1}{r} \frac{\partial \alpha}{\partial \theta} \mathbf{e}_\theta + \frac{1}{r \sin \theta} \frac{\partial \alpha}{\partial \phi} \mathbf{e}_\phi. \qquad \text{(A.8-3)}$$

A.8.2   Divergence of a vector field

In Cartesian coordinates, the gradient of a scalar field α is defined by Eq. (A.5-3), i.e.,

$$\nabla \cdot \mathbf{v} = \frac{\partial v_x}{\partial x} + \frac{\partial v_y}{\partial y} + \frac{\partial v_z}{\partial z}. \qquad \text{(A.8-4)}$$

The divergence of a vector field $\mathbf{v}$ in the cylindrical and spherical coordinates is as follows:

- Cylindrical coordinate system

$$\nabla \cdot \mathbf{v} = \frac{1}{r} \frac{\partial}{\partial r} (r v_r) + \frac{1}{r} \frac{\partial v_\theta}{\partial \theta} + \frac{\partial v_z}{\partial z}. \qquad \text{(A.8-5)}$$

- Spherical coordinate system

$$\nabla \cdot \mathbf{v} = \frac{1}{r^2} \frac{\partial}{\partial r} (r^2 v_r) + \frac{1}{r \sin \theta} \frac{\partial}{\partial \theta} (v_\theta \sin \theta) + \frac{1}{r \sin \theta} \frac{\partial v_\phi}{\partial \phi}. \qquad \text{(A.8-6)}$$

A.8.3   Laplacian of a scalar field

In Cartesian coordinates, the Laplacian of a scalar field α is defined by Eq. (A.5-7), i.e.,

$$\nabla^2 \alpha = \frac{\partial^2 \alpha}{\partial x^2} + \frac{\partial^2 \alpha}{\partial y^2} + \frac{\partial^2 \alpha}{\partial z^2}. \qquad \text{(A.8-7)}$$

The Laplacian of a scalar field α in the cylindrical and spherical coordinates is as follows:

- Cylindrical coordinate system

$$\nabla^2 \alpha = \frac{1}{r} \frac{\partial}{\partial r} \left(r \frac{\partial \alpha}{\partial r} \right) + \frac{1}{r^2} \frac{\partial^2 \alpha}{\partial \theta^2} + \frac{\partial^2 \alpha}{\partial z^2}. \qquad \text{(A.8-8)}$$

- Spherical coordinate system

$$\nabla^2 \alpha = \frac{1}{r^2} \frac{\partial}{\partial r} \left(r^2 \frac{\partial \alpha}{\partial r} \right) + \frac{1}{r^2 \sin \theta} \frac{\partial}{\partial \theta} \left(\sin \theta \frac{\partial \alpha}{\partial \theta} \right) + \frac{1}{r^2 \sin^2 \theta} \frac{\partial^2 \alpha}{\partial \phi^2}.$$
$$\text{(A.8-9)}$$

Appendix B

Constants and Conversion Factors

Physical Constants

$$\text{Gas constant } (R) = \begin{cases} 82.05 \text{ cm}^3 \cdot \text{atm/mol} \cdot \text{K} \\ 0.08205 \text{ m}^3 \cdot \text{atm/kmol} \cdot \text{K} \\ 1.987 \text{ cal/mol} \cdot \text{K} \\ 8.314 \text{ J/mol} \cdot \text{K} \\ 8.314 \times 10^{-6} \text{ MPa} \cdot \text{m}^3/\text{mol} \cdot \text{K} \\ 8.314 \times 10^{-3} \text{ kPa} \cdot \text{m}^3/\text{mol} \cdot \text{K} \\ 8.314 \times 10^{-5} \text{ bar} \cdot \text{m}^3/\text{mol} \cdot \text{K} \\ 8.314 \times 10^{-2} \text{ bar.L/mol} \cdot \text{K} \\ 8.314 \times 10^{-2} \text{ bar} \cdot \text{m}^3/\text{kmol} \cdot \text{K} \\ 83.14 \text{ bar} \cdot \text{cm}^3/\text{mol} \cdot \text{K} \end{cases}$$

$$\text{Acceleration of gravity } (g) = \begin{cases} 9.8067 \text{ m/s}^2 \\ 32.1740 \text{ ft/s}^2 \end{cases}$$

Avogadro's number $\quad 6.0221415 \times 10^{23}$ entities
(atoms or molecules)/mol

Conversion Factors

Density
$1 \text{ kg/m}^3 = 10^{-3} \text{ g/cm}^3 = 10^{-3} \text{ kg/L}$
$1 \text{ kg/m}^3 = 0.06243 \text{ lb/ft}^3$

Energy, Heat, Work
$1 \text{ J} = 1 \text{ W} \cdot \text{s} = 1 \text{ N} \cdot \text{m} = 10^{-3} \text{ kJ}$
$= 10^{-5} \text{ bar} \cdot \text{m}^3 = 10 \text{ bar} \cdot \text{cm}^3$
$1 \text{ cal} = 4.184 \text{ J}$
$1 \text{ kJ} = 2.7778 \times 10^{-4} \text{ kW} \cdot \text{h} = 0.94783 \text{ Btu}$

Heat capacity   1 kJ/kg · K = 0.239 cal/g · K
                  1 kJ/kg · K = 0.239 Btu/lb.°R

Force $1\,N = 1\,kg \cdot m/s^2 = 10^5\,g \cdot cm/s^2$ (dyne)
                  $1\,N = 0.2248\,lbf = 7.23275\,lb \cdot ft/s^2$ (poundals)

Length $1\,m = 100\,cm = 10^6\,\mu m = 10^9\,nm$
                  1 m = 39.370 in = 3.2808 ft

Mass   1 kg = 1000 g
                  1 kg = 2.2046 lb

Power $1\,W = 1\,J/s = 10^{-3}\,kW$
                  1 kW = 3412.2 Btu/h = 1.341 hp

Pressure $1\,Pa = 1\,N/m^2$
                  $1\,kPa = 10^3\,Pa = 10^{-3}\,MPa$
                  $1\,bar = 10^5\,Pa = 100\,kPa = 0.98692\,atm$
                  1 atm = 1.01325 bar = 101.325 kPa = 760 mmHg
                  $1\,atm = 14.696\,lbf/in^2$

Temperature   1 K = 1.8 °R
                  T (°F) = 1.8 T (°C) + 32

Volume $1\,m^3 = 1000\,L$
                  $1\,m^3 = 6.1022 \times 10^4\,in^3 = 35.313\,ft^3 = 264.17\,gal$

Notation

Dimensions are given in terms of mass (M), length (L), time (t), temperature (T), and dimensionless $(—)$. Boldface symbols are vectors or tensors.

A	area, L^2	
A_e	area of entrances and exits, L^2	
A_f	area of fixed surface, L^2	
A_m	area of moving surface, L^2	
a	acceleration, L/t^2	
D	diameter of a cylinder or a sphere, L	
D_h	hydraulic equivalent diameter, L	
E_v	friction loss, ML^2/t^2	
$\mathbf{e}_i$	unit vector in the i-direction, $—$	
$\mathbf{F}$	force, ML/t^2	
F_D	drag force, ML/t^2	
$\mathbf{f}$	body force per unit mass, L/t^2	
f	friction factor, $—$	
$\mathbf{g}$	acceleration of gravity, L/t^2	
$\mathbf{I}$	identity tensor, Eq. (A.4-5)	
K	kinetic energy per unit mass, L^2/t^2	
k_B	Boltzmann constant, ML^2/t^2T	
L	length, L	
M	molecular weight, $M/$ mol	
m	mass, M	

n	outwardly directed unit normal vector, —
P	pressure, M/Lt^2
P_m	mechanical pressure, M/Lt^2
P_t	thermodynamic pressure, M/Lt^2
P^{vap}	vapor (saturation) pressure, M/Lt^2
$\mathcal{P}$	modified pressure, M/Lt^2
$\mathcal{Q}$	volumetric flow rate, L^3/t
R	gas constant (in $P\widetilde{V} = RT$), ML^2/t^2T mol
R	radius of a cylinder or a sphere, L
R	material position vector, L
r	$\sqrt{x^2 + y^2}$, radial coordinate in cylindrical coordinates, L
r	$\sqrt{x^2 + y^2 + z^2}$, radial coordinate in spherical coordinates, L
r	spatial position vector, L
T	absolute temperature, T
T	torque, ML^2/t^2
t	time, t
V	volume, L^3
v	speed, L/t
v	velocity, L/t
W	width, L
W_s	shaft work, ML^2/t^2
$\dot{W}$	power, ML^2/t^3
w	velocity of the surface, L/t
x, y, z	rectangular coordinates, L

Greek symbols

$\dot{\gamma}$	rate of deformation tensor, $1/t$
Δ	difference
δ	film thickness, boundary layer thickness, L
δ_{ij}	Kronecker delta, Eq. (A.3-3), —
ε_{ijk}	alternating unit tensor, Eq. (A.3-5), —
ε	surface roughness of the pipe, L

η	non-Newtonian viscosity, M/Lt	
θ	$\arctan(y/x)$, angle in cylindrical coordinates, —	
θ	$\arctan\left(\sqrt{x^2+y^2}/z\right)$, angle in spherical coordinates, —	
λ	mean free path, L	
$\boldsymbol{\lambda}$	unit normal vector, —	
μ	viscosity, M/Lt	
ν	kinematic viscosity (μ/ρ), L^2/t	
$\boldsymbol{\pi}$	total momentum flux, M/Lt^2	
ρ	density, M/L^3	
τ_{ij}	shear stress (flux of j-momentum in the i-direction), M/Lt^2	
$\boldsymbol{\tau}$	momentum flux vector, M/Lt^2	
ϕ	gravitational potential energy, ML^2/t^2	
ψ	stream function, dimensions depend on the coordinate system	
Ω	angular velocity, t^{-1}	

Overlines

$\sim$ per mole

$\wedge$ per unit mass

$-$ time-averaged quantity

Bracket

$\langle a \rangle$ average value of a

Subscripts

atm	atmospheric
ch	characteristic
in	inlet
max	maximum
out	out
sys	system
w	wall or surface
∞	free-stream

Dimensionless numbers

Kn   Knudsen number

Re   Reynolds number

Re_L Reynolds number based on the plate length

Mathematical operations

D/Dt substantial (material) derivative, t^{-1}

∇ del operator, L^{-1}

$\ln x$ the logarithm of x to the base e

$\log x$ the logarithm of x to the base 10

$\exp x$ e^x, the exponential function of x

$O(\ )$ "of the order of"

Bibliography

Ashare, E., R. B. Bird and J. A. Lescarboura, 1965, "Falling cylinder viscometer for non-Newtonian fluids", *AIChE J.*, **11** (5), 910–916.

Bird, R. B., R. C. Armstrong and O. Hassager, 1987, *Dynamics of Polymeric Liquids, Vol. 1: Fluid Mechanics*, 2nd Edn., Wiley, New York.

Colebrook, C. F., 1939, "Turbulent flow in pipes with particular reference to the transition region between the smooth and rough pipe laws", *J. Inst. Civil Eng.*, **11**, 133–156.

Fredrickson, A. G. and R. B. Bird, 1958, "Non-Newtonian flow in annuli", *Ind. Eng. Chem.*, **50** (3), 347–352.

Hanks, R.W. and K. M. Larsen, 1979, "The flow of power-law non-Newtonian fluids in concentric annuli", *Ind. Eng. Chem. Fundam.*, **18**, 33–35.

Moody, L. F., 1944, "Friction factors for pipe flow", *Trans. ASME*, **66**, 671–681.

Reynolds, O., 1886, "On the theory of lubrication and its application to Mr. Beachamp Tower's experiments", *Philos. Trans. Roy. Soc. Lond.*, **177**, 157–235.

Schrimpf, M., J. Esteban, H. Warmeling, T. Farber, A. Behr and A. J. Vorholt, 2021, "Taylor-Couette reactor: Principles, design, and applications", *AIChE J.*, 67:e17228 (https://doi.org/10.1002/aic.17228).

Whitaker, S., 1988, "Levels of simplification: The use of assumptions, restrictions, and constraints in engineering analysis", *Chem. Eng. Ed.*, **22** (2), 104–108.

Whitaker, S., 1992, *Introduction to Fluid Mechanics*, Krieger Pub. Co., Malabar, Florida.

Index

Printed in the United States
by Baker & Taylor Publisher Services

Printed in the United States
by Baker & Taylor Publisher Services